Edith D. Sylla

PATTERNS OF DISCOVERY

PATTERNS OF DISCOVERY

AN INQUIRY INTO THE CONCEPTUAL FOUNDATIONS OF SCIENCE

BY

NORWOOD RUSSELL HANSON
B.Sc., M.A. (Cantab), D.Phil. (Oxon)
Professor of Philosophy at
Yale University

CAMBRIDGE
AT THE UNIVERSITY PRESS
1965

PUBLISHED BY
THE SYNDICS OF THE CAMBRIDGE UNIVERSITY PRESS
Bentley House, 200 Euston Road, London, N.W. 1
American Branch: 32 East 57th Street, New York, N.Y. 10022
Nigerian Office: P.O. Box 33, Ibadan, Nigeria

First printed 1958
Reprinted 1961
First Paperback edition 1965

First printed in Great Britain at the University Press, Cambridge
Reprinted by offset-litho in the United States of America

TO

FAY

CONTENTS

PREFACE

This book owes so much to so many; yet space permits the mention of so few. But the friendly guidance of H. H. Price, G. Ryle, P. F. Strawson, S. E. Toulmin, A. J. T. D. Wisdom and W. C. Kneale amongst philosophers, and Sir H. Jeffreys, P. A. M. Dirac and N. F. Mott amongst physicists, must be gratefully acknowledged. So must my indebtedness to R. A. Becher of the Cambridge University Press, and to the Master and Fellows of St John's College, Cambridge. Without their encouragement, this work would have expired long since, which, I concede, might have been a more deserving fate for it. To my wife, Fay, the most endearing of critics, I give my grateful thanks—for everything.

N.R.H.

BLOOMINGTON, INDIANA
May 1958

Reference numbers refer to the notes on pp. 176–234

INTRODUCTION

...treat of the network and not of what the network describes.
WITTGENSTEIN[1]

This essay stresses philosophical aspects of microphysical thinking. Although elementary particle theory is much discussed by philosophers of science its features are not understood. Philosophers often refer to the uncertainty relations, the identity principle, the correspondence principle, and to theoretical terms like 'ψ'; and these references enliven discussions of causality, determinism, natural law, hypotheses and probability. Rarely, however, do they square with the concepts employed by the physicist. Why is this? Why is microphysics misrepresented by philosophers?

The reason is simple. They have regarded as paradigms of physical inquiry not unsettled, dynamic, research sciences like microphysics, but finished systems, planetary mechanics, optics, electromagnetism and classical thermodynamics. 'After all', they say, 'when microphysics settles down it will be like these polished systems.' Such a remark constitutes a mistake in any approach to microphysics. If this attitude is accepted, the proper activity for philosophers of physics would then appear to be either (1) to study the logic of the deductive systems which carry the content of microphysics, or (2) to study the statistical methods whereby microphysical theories are built up from repetitive samplings of data. These two approaches may apply to 'classical' disciplines. But these are not research sciences any longer, though they were at one time—a fact that historians and philosophers of science are in danger of forgetting. Now, however, they constitute a different kind of physics altogether. Distinctions which at present apply to them ought to be suspect when transferred to research disciplines: indeed, these distinctions afford an artificial account even of the kinds of activities in which Kepler, Galileo and Newton themselves were actually engaged.

General conceptions formed on the basis of this first mistake should be equally suspect. *Observation*, *evidence*, *facts*; these notions, if drawn from the 'catalogue-sciences' of school and undergraduate text-books, will ill prepare one for understanding the foundations of particle theory. So too with the ideas of *theory*, *hypothesis*, *law*, *causality* and *principle*. In a growing research discipline, inquiry is directed not to rearranging old facts and explanations into more elegant formal patterns, but rather to the discovery of new patterns of explanation. Hence the philosophical flavour of such ideas differs from that presented by science masters, lecturers, and many philosophers of science.

This discrepancy leads to the familiar notion that contemporary physical thought diverges on all matters of principle from the thinking of the past. Superficially, it seems as if there has been a quantum jump in the history of science. The conceptual explosions of 1901, 1905, 1911, 1913, 1915 and 1924–30 may appear to have transformed physics from the careful, cumulative, experimental discipline it was for Faraday and Kelvin into something more speculative, anomalous, and even frightening. The continuity that historians like Tannery, Duhem and Sarton taught us to look for breaks down abruptly when one supposes Einstein, Bohr, Heisenberg and Dirac to be different kinds of thinkers from Galileo, Kepler and Newton. But this is wrong. These are all physicists: that is, natural philosophers seeking explanations of phenomena in ways more similar than the dichotomy 'classical-modern' has led philosophers of science to imagine.

The approach and method of this essay is unusual. I have chosen not to isolate general philosophical issues—the nature of observation, the status of facts, the logic of causality, and the character of physical theory—and use the conclusions of such inquiries as lenses through which to view particle theory. Rather the reverse: the inadequacy of philosophical discussions of these subjects has inclined me to give a different priority. Particle theory will be the lens through which these perennial philosophical problems will be viewed.

The first five chapters serve as accounts of scientific observation, of the interplay between facts and the notations in which they are expressed, of the 'theory-laden' character of causal talk, of the reasoning involved in forming a physical theory and of the function of law statements in physics. These chapters apply to all scientific inquiry, but they are written with the final chapter in mind. Any argument not applicable to microphysics has been held generally suspect; conversely, otherwise sound arguments have been regarded as established if they help one to understand the conceptual basis of elementary particle theory.

The issue is not theory-using, but theory-finding; my concern is not with the testing of hypotheses, but with their discovery. Let us examine not how observation, facts and data are built up into general systems of physical explanation, but how these systems are built into our observations, and our appreciation of facts and data. Only this will make intelligible the disagreements about the interpretation of terms and symbols within quantum theory.

I have not hesitated to refer to events in the history of physics; these will punctuate the other arguments. This comports with my conception of philosophy of science: namely, that profitable philosophical discussion of any science depends on a thorough familiarity with its history and its present state.

CHAPTER I

OBSERVATION

Were the eye not attuned to the Sun,
The Sun could never be seen by it.

GOETHE[1]

A

Consider two microbiologists. They look at a prepared slide; when asked what they see, they may give different answers. One sees in the cell before him a cluster of foreign matter: it is an artefact, a coagulum resulting from inadequate staining techniques. This clot has no more to do with the cell, *in vivo*, than the scars left on it by the archaeologists spade have to do with the original shape of some Grecian urn. The other biologist identifies the clot as a cell organ, a 'Golgi body'. As for techniques, he argues: 'The standard way of detecting a cell organ is by fixing and staining. Why single out this one technique as producing artefacts, while others disclose genuine organs?'

The controversy continues.[2] It involves the whole theory of microscopical technique; nor is it an obviously experimental issue. Yet it affects what scientists say they see. Perhaps there is a sense in which two such observers do not see the same thing, do not begin from the same data, though their eyesight is normal and they are visually aware of the same object.

Imagine these two observing a Protozoon—*Amoeba*. One sees a one-celled animal, the other a non-celled animal. The first sees *Amoeba* in all its analogies with different types of single cells: liver cells, nerve cells, epithelium cells. These have a wall, nucleus, cytoplasm, etc. Within this class *Amoeba* is distinguished only by its independence. The other, however, sees *Amoeba's* homology not with single cells, but with whole animals. Like all animals *Amoeba* ingests its food, digests and assimilates it. It excretes, reproduces and is mobile—more like a complete animal than an individual tissue cell.

This is not an experimental issue, yet it can affect experiment. What either man regards as significant questions or relevant data can be determined by whether he stresses the first or the last term in 'unicellular animal'.[1]

Some philosophers have a formula ready for such situations: 'Of course they see the same thing. They make the same observation since they begin from the same visual data. But they interpret what they see differently. They construe the evidence in different ways.'[2] The task is then to show how these data are moulded by different theories or interpretations or intellectual constructions.

Considerable philosophers have wrestled with this task. But in fact the formula they start from is too simple to allow a grasp of the nature of observation within physics. Perhaps the scientists cited above do not begin their inquiries from the same data, do not make the same observations, do not even see the same thing? Here many concepts run together. We must proceed carefully, for wherever it makes sense to say that two scientists looking at *x* do not see the same thing, there must always be a prior sense in which they do see the same thing. The issue is, then, 'Which of these senses is most illuminating for the understanding of observational physics?'

These biological examples are too complex. Let us consider Johannes Kepler: imagine him on a hill watching the dawn. With him is Tycho Brahe. Kepler regarded the sun as fixed: it was the earth that moved. But Tycho followed Ptolemy and Aristotle in this much at least: the earth was fixed and all other celestial bodies moved around it. *Do Kepler and Tycho see the same thing in the east at dawn?*

We might think this an experimental or observational question, unlike the questions 'Are there Golgi bodies?' and 'Are Protozoa one-celled or non-celled?'. Not so in the sixteenth and seventeenth centuries. Thus Galileo said to the Ptolemaist '...neither Aristotle nor you can prove that the earth is *de facto* the centre of the universe...'.[3] 'Do Kepler and Tycho see the same thing in the east at dawn?' is perhaps not a *de facto* question either, but rather the beginning of an examination of the concepts of seeing and observation.

The resultant discussion might run:

'Yes, they do.'

'No, they don't.'

'Yes, they do!'

'No, they don't!'...

That this is possible suggests that there may be reasons for both contentions.[1] Let us consider some points in support of the affirmative answer.

The physical processes involved when Kepler and Tycho watch the dawn are worth noting. Identical photons are emitted from the sun; these traverse solar space, and our atmosphere. The two astronomers have normal vision; hence these photons pass through the cornea, aqueous humour, iris, lens and vitreous body of their eyes in the same way. Finally their retinas are affected. Similar electro-chemical changes occur in their selenium cells. The same configuration is etched on Kepler's retina as on Tycho's. So they see the same thing.

Locke sometimes spoke of seeing in this way: a man sees the sun if his is a normally-formed retinal picture of the sun. Dr Sir W. Russell Brain speaks of our retinal sensations as indicators and signals. Everything taking place behind the retina is, as he says, 'an intellectual operation based largely on non-visual experience...'.[2] What we *see* are the changes in the *tunica retina*. Dr Ida Mann regards the macula of the eye as itself 'seeing details in bright light', and the rods as 'seeing approaching motor-cars'. Dr Agnes Arber speaks of the eye as itself seeing.[3] Often, talk of seeing can direct attention to the retina. Normal people are distinguished from those for whom no retinal pictures can form: we may say of the former that they can see whilst the latter cannot see. Reporting when a certain red dot can be seen may supply the occulist with direct information about the condition of one's retina.[4]

This need not be pursued, however. These writers speak carelessly: seeing the sun is not seeing retinal pictures of the sun. The retinal images which Kepler and Tycho have are four in number, inverted and quite tiny.[5] Astronomers cannot be referring to these when they say they see the sun. If they are hypnotized, drugged, drunk or distracted they may not see the sun, even though their retinas register its image in exactly the same way as usual.

Seeing is an experience. A retinal reaction is only a physical state—a photochemical excitation. Physiologists have not always appreciated the differences between experiences and physical states.[6] People, not their eyes, see. Cameras, and eye-balls, are blind. Attempts to locate within the organs of sight (or within the neurological reticulum behind the eyes) some nameable called 'seeing' may be dismissed. That Kepler and Tycho do, or do not, see the same thing cannot be supported by reference to the physical states of their retinas, optic nerves or visual cortices: there is more to seeing than meets the eyeball.

Naturally, Tycho and Kepler see the same physical object. They are both visually aware of the sun. If they are put into a dark room and asked to report when they see something—anything at all—they may both report the same object at the same time. Suppose that the only object to be seen is a certain lead cylinder. Both men see the same thing: namely this object—whatever it is. It is just here, however, that the difficulty arises, for while Tycho sees a mere pipe, Kepler will see a telescope, the instrument about which Galileo has written to him.

Unless both are visually aware of the same object there can be nothing of philosophical interest in the question whether or not they see the same thing. Unless they both see the sun in this prior sense our question cannot even strike a spark.

Nonetheless, both Tycho and Kepler have a common visual experience of some sort. This experience perhaps constitutes their seeing the same thing. Indeed, this may be a seeing logically more basic than anything expressed in the pronouncement 'I see the sun' (where each means something different by 'sun'). If what they meant by the word 'sun' were the only clue, then Tycho and Kepler could not be seeing the same thing, even though they were gazing at the same object.

If, however, we ask, not 'Do they see the same thing?' but rather 'What is it that they both see?', an unambiguous answer may be forthcoming. Tycho and Kepler are both aware of a brilliant yellow-white disc in a blue expanse over a green one. Such a 'sense-datum' picture is single and uninverted. To be unaware of it is not to have it. Either it dominates one's visual attention completely or it does not exist.

If Tycho and Kepler are aware of anything visual, it must be of some pattern of colours. What else could it be? We do not touch or hear with our eyes, we only take in light.[1] This private pattern is the same for both observers. Surely if asked to sketch the contents of their visual fields they would both draw a kind of semi-circle on a horizon-line.[2] They say they see the sun. But they do not see every side of the sun at once; so what they really see is discoid to begin with. It is but a visual aspect of the sun. In any single observation the sun is a brilliantly luminescent disc, a penny painted with radium.

So something about their visual experiences at dawn is the same for both: a brilliant yellow-white disc centred between green and blue colour patches. Sketches of what they both see could be identical—congruent. In this sense Tycho and Kepler see the same thing at dawn. The sun appears to them in the same way. The same view, or scene, is presented to them both.

In fact, we often speak in this way. Thus the account of a recent solar eclipse:[1] 'Only a thin crescent remains; white light is now completely obscured; the sky appears a deep blue, almost purple, and the landscape is a monochromatic green...there are the flashes of light on the disc's circumference and now the brilliant crescent to the left....' Newton writes in a similar way in the *Opticks*: 'These Arcs at their first appearance were of a violet and blue Colour, and between them were white Arcs of Circles, which...became a little tinged in their inward Limbs with red and yellow....'[2] Every physicist employs the language of lines, colour patches, appearances, shadows. In so far as two normal observers use this language of the same event, they begin from the same data: they are making the same observation. Differences between them must arise in the interpretations they put on these data.

Thus, to summarize, saying that Kepler and Tycho see the same thing at dawn just because their eyes are similarly affected is an elementary mistake. There is a difference between a physical state and a visual experience. Suppose, however, that it is argued as above—that they see the same thing because they have the same sense-datum experience. Disparities in their accounts arise in *ex post facto* interpretations of what is seen, not in the fundamental visual data. If this is argued, further difficulties soon obtrude.

B

Normal retinas and cameras are impressed similarly by fig. 1.[3] Our visual sense-data will be the same too. If asked to draw what we see, most of us will set out a configuration like fig. 1.

Do we all see the same thing?[4] Some will see a perspex cube viewed from below. Others will see it from above. Still others will see it as a kind of polygonally-cut gem. Some people see only criss-crossed lines in a plane. It may be seen as a block of ice, an aquarium, a wire frame for a kite—or any of a number of other things.

Do we, then, all see the same thing? If we do, how can these differences be accounted for?

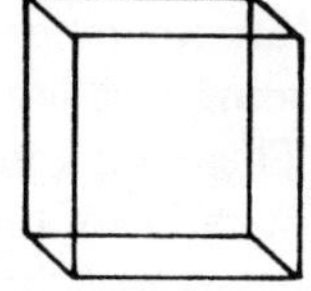

Fig. 1

Here the 'formula' re-enters: 'These are different *interpretations* of what all observers see in common. Retinal reactions to fig. 1 are virtually identical; so too are our visual sense-data, since our drawings of what we see will have the same content. There is no place in the seeing for these differences, so they must lie in the interpretations put on what we see.'

This sounds as if I do two things, not one, when I see boxes and bicycles. Do I put different interpretations on fig. 1 when I see it now as a box from below, and now as a cube from above? I am aware of no such thing. I mean no such thing when I report that the box's perspective has snapped back into the page.[1] If I do not mean this, then the concept of seeing which is natural in this connexion does not designate two diaphanous components, one optical, the other interpretative. Fig. 1 is simply seen now as a box from below, now as a cube from above; one does not first soak up an optical pattern and then clamp an interpretation on it. Kepler and Tycho just see the sun. That is all. That is the way the concept of seeing works in this connexion.

'But', you say, 'seeing fig. 1 first as a box from below, then as a cube from above, involves interpreting the lines differently in each case.' Then for you and me to have a different interpretation of fig. 1 just *is* for us to see something different. This does not mean we see the same thing and then interpret it differently. When I suddenly exclaim 'Eureka—a box from above', I do not refer simply to a different interpretation. (Again, there is a logically prior sense in which seeing fig. 1 as from above and then as from below is seeing the same thing differently, i.e. being aware of the same diagram in different ways. We can refer just to this, but we need not. In this case we do not.)

Besides, the word 'interpretation' is occasionally useful. We know where it applies and where it does not. Thucydides presented the facts objectively; Herodotus put an interpretation on them.

The word does not apply to everything—it has a meaning. Can interpreting always be going on when we see? Sometimes, perhaps, as when the hazy outline of an agricultural machine looms up on a foggy morning and, with effort, we finally identify it. Is this the 'interpretation' which is active when bicycles and boxes are clearly seen? Is it active when the perspective of fig. 1 snaps into reverse? There was a time when Herodotus was half-through with his interpretation of the Graeco-Persian wars. Could there be a time when one is half-through interpreting fig. 1 as a box from above, or as anything else?

'But the interpretation takes very little time—it is instantaneous.' Instantaneous interpretation hails from the Limbo that produced unsensed sensibilia, unconscious inference, incorrigible statements, negative facts and *Objektive*. These are ideas which philosophers force on the world to preserve some pet epistemological or metaphysical theory.

Only in contrast to 'Eureka' situations (like perspective reversals, where one cannot interpret the data) is it clear what is meant by saying that though Thucydides could have put an interpretation on history, he did not. Moreover, whether or not an historian is advancing an interpretation is an empirical question: we know what would count as evidence one way or the other. But whether we are employing an interpretation when we see fig. 1 in a certain way is not empirical. What could count as evidence? In no ordinary sense of 'interpret' do I interpret fig. 1 differently when its perspective reverses for me. If there is some extraordinary sense of word it is not clear, either in ordinary language, or in extraordinary (philosophical) language. To insist that different reactions to fig. 1 *must* lie in the interpretations put on a common visual experience is just to reiterate (without reasons) that the seeing of *x must* be the same for all observers looking at *x*.

'But "I see the figure as a box" means: I am having a particular visual experience which I always have when I interpret the figure as a box, or when I look at a box....' '...if I meant this, I ought to know it. I ought to be able to refer to the experience directly and not only indirectly....'[1]

Ordinary accounts of the experiences appropriate to fig. 1 do not require visual grist going into an intellectual mill: theories and interpretations are 'there' in the seeing from the outset. How can interpretations 'be there' in the seeing? How is it possible to see an object according to an interpretation? 'The question represents it as a queer fact; as if something were being forced into a form it did not really fit. But no squeezing, no forcing took place here.'[1]

Consider now the reversible perspective figures which appear in textbooks on Gestalt psychology: the tea-tray, the shifting (Schröder) staircase, the tunnel. Each of these can be seen as concave, as convex, or as a flat drawing.[2] Do I really see something different each time, or do I only interpret what I see in a different way? To interpret is to think, to do something; seeing is an experiential state.[3] The different ways in which these figures are seen are not due to different thoughts lying behind the visual reactions. What could 'spontaneous' mean if these reactions are not spontaneous? When the staircase 'goes into reverse' it does so spontaneously. One does not think of anything special; one does not think at all. Nor does one interpret. One just sees, now a staircase as from above, now a staircase as from below.

The sun, however, is not an entity with such variable perspective. What has all this to do with suggesting that Tycho and Kepler may see different things in the east at dawn? Certainly the cases are different. But these reversible perspective figures are examples of different things being seen in the same configuration, where this difference is due neither to differing visual pictures, nor to any 'interpretation' superimposed on the sensation.

Fig. 2

Some will see in fig. 2 an old Parisienne, others a young woman (à la Toulouse-Lautrec).[4] All normal retinas 'take' the same picture; and our sense-datum pictures must be the same, for even if you see an old lady and I a young lady, the pictures we draw of what we see may turn out to be geometrically indistinguishable. (Some can see this *only* in one way, not both. This is like the difficulty we have after finding a face in a tree-puzzle; we cannot thereafter see the tree without the face.)

When what is observed is characterized so differently as 'young woman' or 'old woman', is it not natural to say that the observers see different things? Or must 'see different things' mean only 'see different objects'? This is a primary sense of the expression, to be sure. But is there not also a sense in which one who cannot see the young lady in fig. 2 sees something different from me, who sees the young lady? Of course there is.

Similarly, in Köhler's famous drawing of the Goblet-and-Faces[1] we 'take' the same retinal/cortical/sense-datum picture of the configuration; our drawings might be indistinguishable. I see a goblet, however, and you see two men staring at one another. Do we see the same thing? Of course we do. But then again we do not. (The sense in which we *do* see the same thing begins to lose its philosophical interest.)

I draw my goblet. You say 'That's just what I saw, two men in a staring contest'. What steps must be taken to get you to see what I see? When attention shifts from the cup to the faces does one's visual picture change? How? What is it that changes? What could change? Nothing optical or sensational is modified. Yet one sees different things. The organization of what one sees changes.[2]

How does one describe the difference between the *jeune fille* and the *vieille femme* in fig. 2? Perhaps the difference is not describable: it may just show itself.[3] That two observers have not seen the same things in fig. 2 could show itself in their behaviour. What is the difference between us when you see the zebra as black with white stripes and I see it as white with black stripes? Nothing optical. Yet there might be a context (for instance, in the genetics of animal pigmentation), where such a difference could be important.

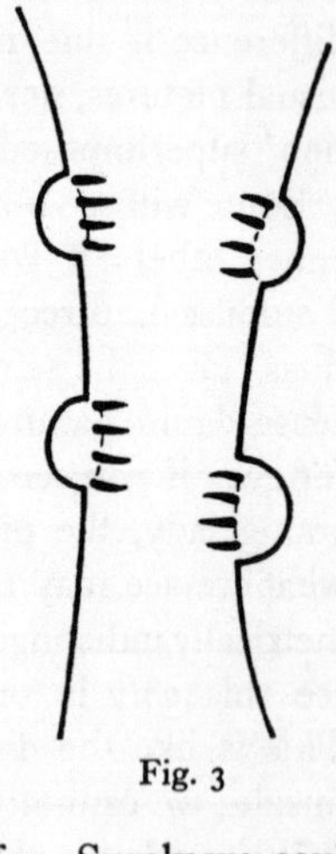

A third group of figures will stress further this organizational element of seeing and observing. They will hint at how much more is involved when Tycho and Kepler witness the dawn than 'the formula' suggests.

Fig. 3

What is portrayed in fig. 3? Your retinas and visual cortices are affected much as mine are; our sense-datum pictures would not differ. Surely we could all produce an accurate sketch of fig. 3. Do we see the same thing?

I see a bear climbing up the other side of a tree. Did the elements 'pull together'/cohere/organize, when you learned this?[1] You might even say with Wittgenstein 'it has not changed, and yet I see it differently...'.[2] Now, does it not have '...a quite particular "organization"'?

Organization is not itself seen as are the lines and colours of a drawing. It is not itself a line, shape, or a colour. It is not an element in the visual field, but rather the way in which elements are appreciated. Again, the plot is not another detail in the story. Nor is the tune just one more note. Yet without plots and tunes details and notes would not hang together. Similarly the organization of fig. 3 is nothing that registers on the retina along with other details. Yet it gives the lines and shapes a pattern. Were this lacking we would be left with nothing but an unintelligible configuration of lines.

How do visual experiences become organized? How is seeing possible?

Consider fig. 4 in the context of fig. 5:

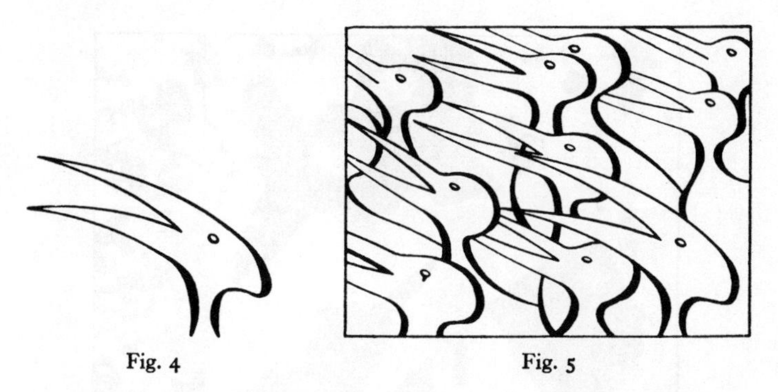

Fig. 4 Fig. 5

The context gives us the clue. Here, some people could not see the figure as an antelope. Could people who had never seen an antelope, but only birds, see an antelope in fig. 4?

In the context of fig. 6 the figure may indeed stand out as an antelope. It might even be urged that the figure seen in fig. 5 has no similarity to the one in fig. 6 although the two are congruent. Could anything be more opposed to a sense-datum account of seeing?

Of a figure similar to the Necker cube (fig. 1) Wittgenstein writes,

'You could imagine [this] appearing in several places in a text-book. In the relevant text something different is in question every time: here a glass cube, there an inverted open box, there a wire frame of that shape, there three boards forming a solid angle. Each time the text supplies the interpretation of the illustration. But we can also see the illustration now as one thing, now as another. So we interpret it, and see it as we interpret it.'[1]

Fig. 6

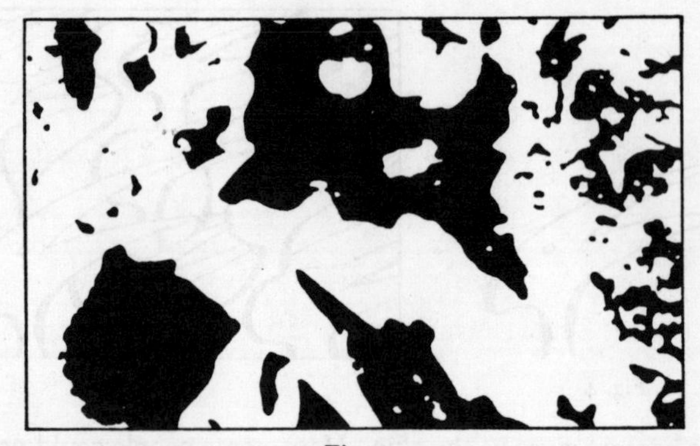

Fig. 7

Consider now the head-and-shoulders in fig. 7:

The upper margin of the picture cuts the brow, thus the top of the head is not shown. The point of the jaw, clean shaven and brightly illuminated, is just above the geometric center of the picture. A white mantle...covers the right shoulder. The right upper sleeve is exposed as the rather black area at the lower left. The hair and beard are after the manner of a late mediaeval representation of Christ.[2]

The appropriate aspect of the illustration is brought out by the verbal context in which it appears. It is not an illustration of anything determinate unless it appears in some such context. In the same way, I must talk and gesture around fig. 4 to get you to see the antelope when only the bird has revealed itself. I must provide a context. The context is part of the illustration itself.

Such a context, however, need not be set out explicitly. Often it is 'built into' thinking, imagining and picturing. We are set[1] to appreciate the visual aspect of things in certain ways. Elements in our experience do not cluster at random.

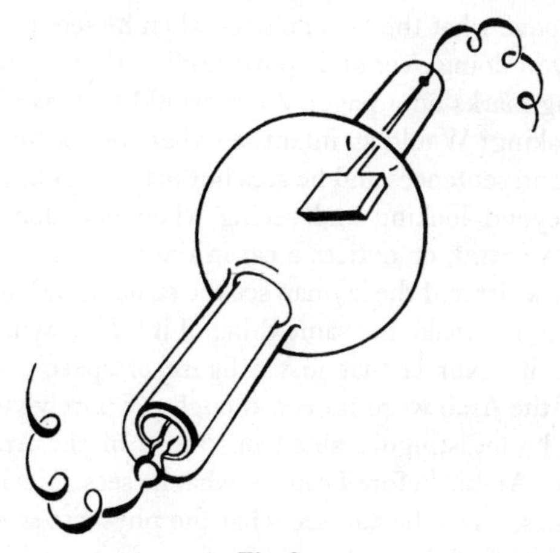

Fig. 8

A trained physicist could see one thing in fig. 8: an X-ray tube viewed from the cathode. Would Sir Lawrence Bragg and an Eskimo baby see the same thing when looking at an X-ray tube? Yes, and no. Yes—they are visually aware of the same object. No—the *ways* in which they are visually aware are profoundly different. Seeing is not only the having of a visual experience; it is also the way in which the visual experience is had.

At school the physicist had gazed at this glass-and-metal instrument. Returning now, after years in University and research, his eye lights upon the same object once again. Does he see the same thing now as he did then? Now he sees the instrument in terms of electrical circuit theory, thermodynamic theory, the theories of metal and glass structure, thermionic emission, optical transmission, refraction, diffraction, atomic theory, quantum theory and special relativity.

Contrast the freshman's view of college with that of his ancient tutor. Compare a man's first glance at the motor of his car with a similar glance ten exasperating years later.

'Granted, one learns all these things', it may be countered, 'but it all figures in the interpretation the physicist puts on what he sees. Though the layman sees exactly what the physicist sees, he cannot interpret it in the same way because he has not learned so much.'

Is the physicist doing more than just seeing? No; he does nothing over and above what the layman does when he sees an X-ray tube. What are you doing over and above reading these words? Are you interpreting marks on a page? When would this ever be a natural way of speaking? Would an infant see what you see here, when you see words and sentences and he sees but marks and lines? One does nothing beyond looking and seeing when one dodges bicycles, glances at a friend, or notices a cat in the garden.

'The physicist and the layman see the same thing', it is objected, 'but they do not make the same thing of it.' The layman can make nothing of it. Nor is that just a figure of speech. I can make nothing of the Arab word for *cat*, though my purely visual impressions may be indistinguishable from those of the Arab who can. I must learn Arabic before I can see what he sees. The layman must learn physics before he can see what the physicist sees.

If one must find a paradigm case of seeing it would be better to regard as such not the visual apprehension of colour patches but things like seeing what time it is, seeing what key a piece of music is written in, and seeing whether a wound is septic.[1]

Pierre Duhem writes:

> Enter a laboratory; approach the table crowded with an assortment of apparatus, an electric cell, silk-covered copper wire, small cups of mercury, spools, a mirror mounted on an iron bar; the experimenter is inserting into small openings the metal ends of ebony-headed pins; the iron oscillates, and the mirror attached to it throws a luminous band upon a celluloid scale; the forward-backward motion of this spot enables the physicist to observe the minute oscillations of the iron bar. But ask him what he is doing. Will he answer 'I am studying the oscillations of an iron bar which carries a mirror'? No, he will say that he is measuring the electric resistance of the spools. If you are astonished, if you ask

him what his words mean, what relation they have with the phenomena he has been observing and which you have noted at the same time as he, he will answer that your question requires a long explanation and that you should take a course in electricity.[1]

The visitor must learn some physics before he can see what the physicist sees. Only then will the context throw into relief those features of the objects before him which the physicist sees as indicating resistance.

This obtains in all seeing. Attention is rarely directed to the space between the leaves of a tree, save when a Keats brings it to our notice.[2] (Consider also what was involved in Crusoe's seeing a vacant space in the sand as a footprint.) Our attention most naturally rests on objects and events which dominate the visual field. What a blooming, buzzing, undifferentiated confusion visual life would be if we all arose tomorrow without attention capable of dwelling only on what had heretofore been overlooked.[3]

The infant and the layman can see: they are not blind. But they cannot see what the physicist sees; they are blind to what he sees.[4] We may not hear that the oboe is out of tune, though this will be painfully obvious to the trained musician. (Who, incidentally, will not hear the tones and *interpret* them as being out of tune, but will simply hear the oboe to be out of tune.[5] We simply see what time it is; the surgeon simply sees a wound to be septic; the physicist sees the X-ray tube's anode overheating.) The elements of the visitor's visual field, though identical with those of the physicist, are not organized for him as for the physicist; the same lines, colours, shapes are apprehended by both, but not in the same way. There are indefinitely many ways in which a constellation of lines, shapes, patches, may be seen. *Why* a visual pattern is seen differently is a question for psychology, but *that* it may be seen differently is important in any examination of the concepts of seeing and observation. Here, as Wittgenstein might have said, the psychological is a symbol of the logical.

You see a bird, I see an antelope; the physicist sees an X-ray tube, the child a complicated lamp bulb; the microscopist sees coelenterate mesoglea, his new student sees only a gooey, formless stuff. Tycho and Simplicius see a mobile sun, Kepler and Galileo see a static sun.[6]

It may be objected, 'Everyone, whatever his state of knowledge, will see fig. 1 as a box or cube, viewed as from above or as from below'. True; almost everyone, child, layman, physicist, will see the figure as box-like one way or another. But could such observations be made by people ignorant of the construction of box-like objects? No. This objection only shows that most of us—the blind, babies, and dimwits excluded—have learned enough to be able to see this figure as a three-dimensional box. This reveals something about the sense in which Simplicius and Galileo do see the same thing (which I have never denied): they both see a brilliant heavenly body. The schoolboy and the physicist both see that the X-ray tube will smash if dropped. Examining how observers see different things in *x* marks something important about their seeing the same thing when looking at *x*. If seeing different things involves having different knowledge and theories about *x*, then perhaps the sense in which they see the same thing involves their sharing knowledge and theories about *x*. Bragg and the baby share no knowledge of X-ray tubes. They see the same thing only in that if they are looking at *x* they are both having some visual experience of it. Kepler and Tycho agree on more: they see the same thing in a stronger sense. Their visual fields are organized in much the same way. Neither sees the sun about to break out in a grin, or about to crack into ice cubes. (The baby is not 'set' even against these eventualities.) Most people today see the same thing at dawn in an even stronger sense: we share much knowledge of the sun. Hence Tycho and Kepler see different things, and yet they see the same thing. That these things can be said depends on their knowledge, experience, and theories.

Kepler and Tycho are to the sun as we are to fig. 4, when I see the bird and you see only the antelope. The elements of their experiences are identical; but their conceptual organization is vastly different. Can their visual fields have a different organization? Then they can see different things in the east at dawn.

It is the sense in which Tycho and Kepler do not observe the same thing which must be grasped if one is to understand disagreements within microphysics. Fundamental physics is primarily a search for intelligibility—it is philosophy of matter. Only secondarily is it a search for objects and facts (though the two endeavours are as hand and glove). Microphysicists seek new modes of conceptual organization. If that can be done the finding of new entities will follow. Gold is rarely discovered by one who has not got the lay of the land.

To say that Tycho and Kepler, Simplicius and Galileo, Hooke and Newton, Priestley and Lavoisier, Soddy and Einstein, De Broglie and Born, Heisenberg and Bohm all make the same observations but use them differently is too easy.[1] It does not explain controversy in research science. Were there no sense in which they were different observations they could not be used differently. This may perplex some: that researchers sometimes do not appreciate data in the same way is a serious matter. It is important to realize, however, that sorting out differences about data, evidence, observation, may require more than simply gesturing at observable objects. It may require a comprehensive reappraisal of one's subject matter. This may be difficult, but it should not obscure the fact that nothing less than this may do.

C

There is a sense, then, in which seeing is a 'theory-laden' undertaking. Observation of *x* is shaped by prior knowledge of *x*. Another influence on observations rests in the language or notation used to express what we know, and without which there would be little we could recognize as knowledge. This will be examined.[2]

I do not mean to identify seeing with *seeing as*. Seeing an X-ray tube is not seeing a glass-and-metal object as an X-ray tube.[3] However, seeing an antelope and seeing an object as an antelope have much in common. Something of the concept of seeing can be discerned from tracing uses of 'seeing...as...'. Wittgenstein is reluctant[4] to concede this, but his reasons are not clear to me. On the contrary, the logic of 'seeing as' seems to illuminate the general perceptual case.[5] Consider again the footprint in the sand. Here all the organizational features of *seeing as* stand out clearly, in the absence of an '*object*'. One can even imagine cases where 'He sees it as a footprint' would be a way of referring to another's apprehension of what actually is a footprint. So, while I do not identify, for example, Hamlet's seeing of a camel in the clouds with his seeing of Yorick's skull, there is still something to be learned about the latter from noting what is at work in the former.

There is, however, a further element in seeing and observation.

If the label 'seeing as' has drawn out certain features of these concepts, 'seeing that...' may bring out more. Seeing a bear in fig. 3 was to see that were the 'tree' circled we should come up behind the beast. Seeing the dawn was for Tycho and Simplicius to see that the earth's brilliant satellite was beginning its diurnal circuit around us, while for Kepler and Galileo it was to see that the earth was spinning them back into the light of our local star. Let us examine 'seeing that' in these examples. It may be the logical element which connects observing with our knowledge, and with our language.

Of course there are cases where the data are confused and where we may have no clue to guide us. In microscopy one often reports sensations in a phenomenal, lustreless way: 'it is green in this light; darkened areas mark the broad end....' So too the physicist may say: 'the needle oscillates, and there is a faint streak near the neon parabola. Scintillations appear on the periphery of the cathodescope....' To deny that these are genuine cases of seeing, even observing, would be unsound, just as is the suggestion that they are the *only* genuine cases of seeing.

These examples are, however, overstressed. The language of shapes, colour patches, oscillations and pointer-readings is appropriate to the unsettled experimental situation, where confusion and even conceptual muddle may dominate. The observer may not know what he is seeing: he aims only to get his observations to cohere against a background of established knowledge. This seeing is the goal of observation. It is in these terms, and not in terms of 'phenomenal' seeing, that new inquiry proceeds. Every physicist forced to observe his data as in an oculist's office finds himself in a special, unusual situation. He is obliged to forget what he knows and to watch events like a child. These are non-typical cases, however spectacular they may sometimes be.

First registering observations and then casting about for knowledge of them gives a simple model of how the mind and the eye fit together. The relationship between seeing and the corpus of our knowledge, however, is not a simple one.

What is it to see boxes, staircases, birds, antelopes, bears, goblets, X-ray tubes? It is (at least) to have knowledge of certain sorts. (Robots and electric eyes are blind, however efficiently they react to light. Cameras cannot see.) It is to see that, were certain things done to objects before our eyes, other things would result. How should we regard a man's report that he sees x if we know him to be ignorant of all x-ish things? Precisely as we would regard a four-year-old's report that he sees a meson shower. 'Smith sees x' suggests that Smith could specify some things pertinent to x. To see an X-ray tube is at least to see that, were it dropped on stone, it would smash. To see a goblet is to see something with concave interior. We may be wrong, but not always—not even usually. Besides, deceptions proceed in terms of what is normal, ordinary. Because the world is not a cluster of conjurer's tricks, conjurers can exist. Because the logic of 'seeing that' is an intimate part of the concept of seeing, we sometimes rub our eyes at illusions.

'Seeing as' and 'seeing that' are not components of seeing, as rods and bearings are parts of motors: seeing is not composite. Still, one *can* ask logical questions. What must have occurred, for instance, for us to describe a man as having found a collar stud, or as having seen a bacillus? Unless he had had a visual sensation and knew what a bacillus was (and looked like) we would not say that he had seen a bacillus, except in the sense in which an infant could see a bacillus. 'Seeing as' and 'seeing that', then, are not psychological components of seeing. They are logically distinguishable elements in seeing-talk, in our concept of seeing.

To see fig. 1 as a transparent box, an ice-cube, or a block of glass is to see that it is six-faced, twelve-edged, eight-cornered. Its corners are solid right angles; if constructed it would be of rigid, or semi-rigid material, not of liquescent or gaseous stuff like oil, vapour or flames. It would be tangible. It would take up space in an exclusive way, being locatable here, there, but at least somewhere. Nor would it cease to exist when we blinked. Seeing it as a cube is just to see that all these things would obtain.

This is knowledge: it is knowing what kind of a thing 'box' or 'cube' denotes and something about what materials can make up such an entity. 'Transparent box' or 'glass cube' would not express what was seen were any of these further considerations denied. Seeing a bird in the sky involves seeing that it will not suddenly do vertical snap rolls; and this is more than marks the retina. We could be wrong. But to see a bird, even momentarily, is to see it in all these connexions. As Wisdom would say, every perception involves an aetiology and a prognosis.[1]

Sense-datum theorists stress how we can go wrong in our observations, as when we call aeroplanes 'birds'. Thus they seek what we are right about, even in these cases. Preoccupation with this problem obscures another one, namely, that of describing what is involved when we are right about what we say we see; and after all this happens very often. His preoccupation with mistakes leads the phenomenalist to portray a world in which we are usually deceived; but the world of physics is not like that. Were a physicist in an ordinary laboratory situation to react to his visual environment with purely sense-datum responses—as does the infant or the idiot—we would think him out of his mind. We would think him *not* to be seeing what was around him.

'Seeing that' threads knowledge into our seeing; it saves us from re-identifying everything that meets our eye; it allows physicists to observe new data as physicists, and not as cameras. We do not ask 'What's that?' of every passing bicycle. The knowledge is there in the seeing and not an adjunct of it. (The pattern of threads is there in the cloth and not tacked on to it by ancillary operations.) We rarely catch ourselves tacking knowledge on to what meets the eye. Seeing this page as having an opposite side requires no squeezing or forcing, yet nothing optical guarantees that when you turn the sheet it will not cease to exist. This is but another way of saying that ordinary seeing is corrigible, which everybody would happily concede. The search for incorrigible seeing has sometimes led some philosophers to deny that anything less than the incorrigible is seeing at all.

Seeing an object x is to see that it may behave in the ways we know x's do behave: if the object's behaviour does not accord with what we expect of x's we may be blocked from seeing it as a straightforward x any longer. Now we rarely see dolphin as fish, the earth as flat, the heavens as an inverted bowl or the sun as our satellite. '...what I perceive as the dawning of an aspect is not a property of the object, but an internal relation between it and other objects.'[1] To see in fig. 8 an X-ray tube is to see that a photo-sensitive plate placed below it will be irradiated. It is to see that the target will get extremely hot, and as it has no water-jacket it must be made of metal with a high melting-point—molybdenum or tungsten. It is to see that at high voltages green fluorescence will appear at the anode. Could a physicist see an X-ray tube without seeing that these other things would obtain? Could one see something as an incandescent light bulb and fail to see that it is the wire filament which 'lights up' to a white heat? The answer may sometimes be 'yes', but this only indicates that different things can be meant by 'X-ray tube' and 'incandescent bulb'. Two people confronted with an *x* may mean different things by *x*. Must their saying 'I see *x*' mean that they see the same thing? A child could parrot 'X-ray tube', or 'Kentucky' or 'Winston', when confronted with the figure above, but he would not see that these other things followed. And this is what the physicist does see.

If in the brilliant disc of which he is visually aware Tycho sees only the sun, then he cannot but see that it is a body which will behave in characteristically 'Tychonic' ways. These serve as the foundation for Tycho's general geocentric-geostatic theories about the sun. They are not imposed on his visual impressions as a tandem interpretation: they are 'there' in the seeing. (So too the interpretation of a piece of music is there in the music. Where else could it be? It is not something superimposed upon pure, unadulterated sound.)

Similarly we see fig. 1 as from underneath, as from above, or as a diagram of a rat maze or a gem-cutting project. However construed, the construing is there in the seeing. One is tempted to say 'the construing *is* the seeing'. The thread and its arrangement *is* the fabric, the sound and its composition *is* the music, the colour and its disposition *is* the painting. There are not two operations involved in my seeing fig. 1 as an ice-cube; I simply see it as an ice-cube. Analogously, the physicist sees an X-ray tube, not by first soaking up reflected light and then clamping on interpretations, but just as you see this page before you.

Tycho sees the sun beginning its journey from horizon to horizon. He sees that from some celestial vantage point the sun (carrying with it the moon and planets) could be watched circling our fixed earth. Watching the sun at dawn through Tychonic spectacles would be to see it in something like this way.

Kepler's visual field, however, has a different conceptual organization. Yet a drawing of what he sees at dawn could be a drawing of exactly what Tycho saw,[1] and could be recognized as such by Tycho. But Kepler will see the horizon dipping, or turning away, from our fixed local star. The shift from sunrise to horizon-turn is analogous to the shift-of-aspect phenomena already considered; it is occasioned by differences between what Tycho and Kepler think they know.

These logical features of the concept of seeing are inextricable and indispensable to observation in research physics. Why indispensable? That men do see in a way that permits analysis into 'seeing as' and 'seeing that' factors is one thing; 'indispensable', however, suggests that the world must be seen thus. This is a stronger claim, requiring a stronger argument. Let us put it differently: that observation in physics is not an encounter with unfamiliar and unconnected flashes, sounds and bumps, but rather a calculated meeting with these as flashes, sounds and bumps of a particular kind—this might figure in an account of what observation is. It would not secure the point that observation could not be otherwise. This latter type of argument is now required: it must establish that an alternative account would be not merely false, but absurd. To this I now turn.

D

Fortunately, we do not see the sun and the moon as we see the points of colour and light in the oculist's office; nor does the physicist see his laboratory equipment, his desk, or his hands in the baffled way that he may view a cloud-chamber photograph or an oscillograph pattern. In most cases we could give further information about what sort of thing we see. This might be expressed in a list: for instance, that x would break if dropped, that x is hollow, and so on.

To see fig. 3 as a bear on a tree is to see that further observations are possible; we can imagine the bear as viewed from the side or from behind. Indeed, seeing fig. 3 as a bear is just to have seen that these other views could all be simultaneous. It is also to see that certain observations are not possible: for example, the bear cannot be waving one paw in the air, nor be dangling one foot. This too is 'there' in the seeing.

'Is it a question of both seeing and thinking? or an amalgam of the two, as I should almost like to say?'[1] Whatever one would like to say, there is more to seeing fig. 3 as a bear, than optics, photochemistry or phenomenalism can explain.[2]

Notice a logical feature: 'see that' and 'seeing that' are always followed by 'sentential' clauses. The addition of but an initial capital letter and a full stop sets them up as independent sentences. One can see an ice-cube, or see a kite as a bird. One cannot see that an ice-cube, nor see that a bird. Nor is this due to limitations of vision. Rather, one may see that *ice-cubes can melt*; that *birds have 'hollow' bones*. Tycho and Simplicius see that *the universe is geocentric*; Kepler and Galileo see that *it is heliocentric*. The physicist sees that *anode-fluorescence will appear in an X-ray tube at high voltages*. The phrases in italics are complete sentential units.

Pictures and statements differ in logical type, and the steps between visual pictures and the statements of what is seen are many and intricate. Our visual consciousness is dominated by pictures; scientific knowledge, however, is primarily linguistic. Seeing is, as I should almost like to say, an amalgam of the two—pictures and language. At the least, the concept of seeing embraces the concepts of visual sensation and of knowledge.[1]

The gap between pictures and language locates the logical function of 'seeing that'. For vision is essentially pictorial, knowledge fundamentally linguistic. Both vision and knowledge are indispensable elements in seeing; but differences between pictorial and linguistic representation may mark differences between the optical and conceptual features of seeing. This may illuminate what 'seeing that' consists in.

Not all the elements of statement correspond to the elements of pictures: only someone who misunderstood the uses of language would expect otherwise.[2] There is a 'linguistic' factor in seeing, although there is nothing linguistic about what forms in the eye, or in the mind's eye. Unless there were this linguistic element, nothing we ever observed could have relevance for our knowledge. We could not speak of significant observations: nothing seen would make sense, and microscopy would only be a kind of kaleidoscopy. For what is it for things to make sense other than for descriptions of them to be composed of meaningful sentences?

We must explore the gulf between pictures and language, between sketching and describing, drawing and reporting. Only by showing how picturing and speaking are different can one suggest how 'seeing that' may bring them together; and brought together they must be if observations are to be *significant* or *noteworthy*.

Knowledge here is of what there is, as factually expressed in books, reports, and essays. How to do things is not our concern. I know how to whistle; but could I express that knowledge in language? Could I describe the taste of salt, even though I know perfectly well how salt tastes? I know how to control a parachute—much of that knowledge was imparted in lectures and drills, but an essential part of it was not *imparted* at all; it was 'got on the spot'. Physicists rely on 'know-how', on the 'feel' of things and the 'look' of a situation, for these control the direction of research. Such imponderables, however, rarely affect the corpus of physical truths. It is not Galileo's insight, Newton's genius and Einstein's imagination which have *per se* changed our knowledge of what there is: it is the true things they have said. 'Physical knowledge', therefore, will mean 'what is reportable in the texts, reports and discussions of physics.' We are concerned with *savoir*, not *savoir faire*.[1]

The 'foundation' of the language of physics, the part closest to mere sensation, is a series of statements. Statements are true or false. Pictures are not at all like statements: they are neither true nor false; retinal, cortical, or sense-datum pictures are neither true nor false. Yet what we see can determine whether statements like 'The sun is above the horizon' and 'The cube is transparent', are true or false. Our visual sensations may be 'set' by language forms; how else could they be appreciated in terms of what we know? Until they *are* so appreciated they do not constitute observation: they are more like the buzzing confusion of fainting or the vacant vista of aimless staring through a railway window.[2] Knowledge of the world is not a *montage* of sticks, stones, colour patches and noises, but a system of propositions.

Fig. 8, p. 15, asserts nothing. It could be inaccurate, but it could not be a lie. This is the wedge between pictures and language.

Significance, *relevance*—these notions depend on what we already know. Objects, events, pictures, are not intrinsically significant or relevant. If seeing were just an optical-chemical process, then nothing we saw would ever be relevant to what we know, and nothing known could have significance for what we see. Visual life would be unintelligible; intellectual life would lack a visual aspect. Man would be a blind computer harnessed to a brainless photoplate.[3]

Pictures often copy originals. All the elements of a copy, however, have the same kind of function. The lines depict elements in the original. The arrangement of the copy's elements shows the disposition of elements in the original. Copy and original are of the same logical type; you and your reflexion are of the same type. Similarly, language might copy what it describes.[1]

Consider fig. 3 alongside 'The bear is on the tree'. The picture contains a bear-element and a tree-element. If it is true to life, then in the original there is a bear and a tree. If the sentence is true *of* life, then (just as it contains 'bear' and 'tree') the situation it describes contains a bear and a tree. The picture combines its elements, it mirrors the actual relation of the bear and the tree. The sentence likewise conjoins 'bear' and 'tree' in the schema 'The —— is on the ——'. This verbal relation signifies the actual relation of the real bear and the real tree. Both picture and sentence are true copies: they contain nothing the original lacks, and lack nothing the original contains. The elements of the picture stand for (represent) elements of the original: so do 'bear' and 'tree'. This is more apparent when expressed symbolically as $b\ R\ t$, where b = bear, t = tree and R = the relation of being on.

By the arrangement of their elements these copies show the arrangement in the original situation. Thus fig. 3, 'The bear is on the tree', and 'bRt' show what obtains with the real bear and the real tree; while 'The tree is on the bear', and 'tRb', and a certain obvious cluster of lines do not show what actually obtains.

The copy is of the same type as the original. We can sketch the bear's teeth, but not his growl, any more than we could see the growl of the original bear. Leonardo could draw Mona Lisa's smile, but not her laugh. Language, however, is more versatile. Here is a dissimilarity between picturing and asserting which will grow to fracture the account once tendered by Wittgenstein, Russell and Wisdom. Language can encapsulate scenes and sounds, teeth and growls, smiles and laughs; a picture, or a gramophone, can do one or the other, but not both. Pictures and recordings stand for things by possessing certain properties of the original itself. Images, reflections, pictures and maps duplicate the spatial properties of what they image, reflect, picture or map; gramophone recordings duplicate audio-temporal properties. Sentences are not like this. They do not stand for things in virtue of possessing properties of the original; they do not *stand for* anything. They can state what is, or could be, the case. They can be used to make assertions, convey descriptions, supply narratives, accounts, etc., none of which depend on the possession of some property in common with what the statement is about. We need not write 'THE BEAR is bigger than ITS CUB' to show our meaning.

Images, reflexions, pictures and maps in fact copy originals with different degrees of strictness. A reflexion of King's Parade does not copy in the same sense that a charcoal sketch does, and both differ from the representation of 'K.P.' on a map of Cambridge and from a town-planner's drawing. The more like a reflection a map becomes, the less useful it is as a map. Drawings are less like copies of originals than are photographs. Of a roughly sketched ursoid shape one says either 'That's a bear' or 'That's supposed to be a bear'. Similarly with maps. Of a certain dot on the map one says either 'This is Cambridge' or 'This stands for Cambridge'.

Language copies least of all. There are exceptional words like 'buzz', 'tinkle' and 'toot', but they only demonstrate how conventional our languages and notations are. Nothing about 'bear' looks like a bear; nothing in the sound of 'bear' resembles a growl. That b-e-a-r can refer to bears is due to a convention which co-ordinates the word with the object. There is nothing dangerous about a red flag, yet it is a signal for danger. Of fig. 3 we might say 'There is a bear'. We would never say this of the word 'bear'. At the cinema we say 'It's a bear', or 'There's K.P.'—not 'That stands for a bear', or 'That denotes K.P.' It is words that denote; but they are rarely like what they denote.

Sentences do not show, for example, bears climbing trees, but they can state that bears climb trees. Showing the sun climbing into the sky consists in representing sun and sky and arranging them appropriately. Stating that the sun is climbing into the sky consists in referring to the sun and then characterizing it as climbing into the sky. The differences between representing and referring, between arranging and characterizing—these are the differences between picturing and language-using.

These differences exist undiminished between visual sense-data and basic sentences. Early logical constructionists were inattentive to the difficulties in fitting visual sense-data to basic sentences. Had they heeded the differences between pictures and maps, they might have detected greater differences still between pictures and language. One's visual awareness of a brown ursoid patch is logically just as remote from the utterance '(I am aware of a) brown, ursoid patch now', as with any of the pictures and sentences we have considered. The picture is of *x*; the statement is to the effect that *x*. The picture shows *x*; the statement refers to and describes *x*. The gap between pictures and language is not closed one millimetre by focusing on sense-data and basic sentences.

The prehistory of languages need not detain us. The issue concerns differences between *our* languages and *our* pictures, and not the smallness of those differences at certain historical times. Wittgenstein is misleading about this: '...and from [hieroglyphic writing] came the alphabet without the essence of representation being lost.'[1] This strengthened the picture theory of meaning, a truth-functional account of language and a theory of atomic sentences. But unless the essence of representation had been lost, languages could not be used in speaking the truth, telling lies, referring and characterizing.

Not all elements of a sentence do the same work. All the elements of pictures, however, just represent.[2] A picture of the dawn could be cut into small pictures, but sentences like 'The sun is on the horizon' and 'I perceive a solaroid patch now' cannot be cut into small sentences. All the elements of the picture show something; none of the elements of the sentences state anything. 'Bear!' may serve as a statement, as may 'Tree!' from the woodcutter, or 'Sun!' in eclipse-observations. But 'the', 'is' and 'on' are not likely ever to behave as statements.

Pictures are of the picturable. Recordings are of the recordable. You cannot play a smile or a wink on the gramophone. But language is more versatile: we can describe odours, sounds, feels, looks, smiles and winks. This freedom makes type-mistakes possible: for example, 'They found his pituitary but not his mind', 'We surveyed his retina but could not find his sight'. Only when we are free from the natural limitations of pictures and recordings can such errors occur. They are just possible in maps; of the hammer and sickle which signifies Russia on a school map, for instance, a child might ask 'How many miles long is the sickle?'. Maps with their partially conventional characters must be read (as pictures and photographs need not be); yet they must copy.

There is a corresponding gap between visual pictures and what we know. Seeing bridges this, for while seeing is at the least a 'visual copying' of objects, it is also more than that. It is a certain sort of seeing of objects: seeing that if x were done to them y would follow. This fact got lost in all the talk about knowledge arising from sense experience, memory, association and correlation. Memorizing, associating, correlating and comparing mental pictures may be undertaken *ad indefinitum* without one step having been taken towards scientific knowledge, that is, propositions known to be true. How long must one shuffle photographs, diagrams and sketches of antelopes before the statement 'antelopes are ungulates' springs forth?

When language and notation are ignored in studies of observation, physics is represented as resting on sensation and low-grade experiment. It is described as repetitious, monotonous concatenation of spectacular sensations, and of school-laboratory experiments. But physical science is not just a systematic exposure of the senses to the world; it is also a way of thinking about the world, a way of forming conceptions. The paradigm observer is not the man who sees and reports what all normal observers see and report, but the man who sees in familiar objects what no one else has seen before.[1]

CHAPTER II

FACTS

A

For 'tough-minded' philosophers, observation is just opening one's eyes and looking. Facts are simply the things that happen; hard, sheer, plain and unvarnished.[1]

We may all speak as we please. A conceptual difficulty, however, lies beneath this talk, one related to the views about seeing and observing which we have already examined. Most of us talk of observing facts, looking at them, collecting them, etc. What is observation of a fact?[2] What would one look like? In what could it be collected? I can photograph an object, an event, or even a situation. What would a photograph of a fact be a photograph of? Is a sketch of the dawn a sketch of the *fact* of which Tycho, Simplicius, Kepler and Galileo were aware? Asking the question suffices. Facts are not picturable, observable entities. Seeing the sun on the horizon involves more than soaking up optical sensibilia, as we have said; but not *so* much more that 'seeing the fact that the sun is on the horizon' fails to jar the ear.

At midnight it is not a fact that the sun is on the horizon. What we see at dawn is therefore relevant to its being a fact that the sun is on the horizon. Still, what we see does not constitute this fact.

What would an inexpressible fact be like? Not a complicated, incompletely understood fact, but one 'inexpressible in principle', a fact which constitutionally resists articulation? When would we be referred to a fact with the aside that it must always elude linguistic expression? To what would we have been directed? Unknown facts of course elude expression. We do not know next year's football scores. Claudius Ptolemy could not express in the second century A.D. what were facts for Galileo fifteen centuries later. But could a fact discovered (and spoken of as known) elude linguistic expression? Could one know facts for which no expression was available, and what sense is there in even speaking of unknowable facts? *Scire est dicere posse.*

This much proves little. The music we play and sing is written in score. This does not prove that music must be writable in score form. Or does it?

Imagine us on Castle Hill this morning, at dawn. I say: 'It is a fact that the sun is on the horizon, whether or not there is a language in which to say so. And this grass at my feet is green. Whether everyone (or no one) knows these things, they remain stubborn, brute facts.'

'The sun is round' states a fact invariant to context: the fact is that the sun is round, however we express this. Is this so also with 'The sun is bright yellow'? A difference appears. Were a man blind to yellow, his statements about the sun's colour would not affect the facts: the sun is yellow. But suppose a super-thermo-nuclear bomb exploded. Its radioactive products destroy retinal sensitivity to yellow and green light: all eyes become electrochemically inert to radiation in the Å 5000–Å 7000 range. Everyone would be colour blind to green and yellow. This sounds odd already. Men are not blind to ultra-violet or infra-red. Only if these 'colours' could be seen would anyone be blind to them. Are we blind to radio waves, sound waves, shock waves? No, they are not visible. ('We are blind to *x*' is not equivalent to '*x* is invisible'.) Under these conditions we would not be blind to green and yellow. This radiation would be invisible—as with ultra-violet and infra-red now. What, then, would be the force of 'This is yellow' or 'That is green'? Exactly that of 'This is ultra-violet', 'That is infra-red': the catastrophe would shift the significance of these expressions from the domain of colour perception to one of pointer-readings, photocell reactions, chemical changes.

Try another tack. We say 'The sun is yellow', 'The grass is green', 'Sugar is sweet', 'Bears are furry'. In this adjectival idiom, yellowness, greenness and sweetness are properties which inhere (passively) in the sun, grass, sugar and bears. They are built into the objects of which we speak. Now convey such information by means of verbs, as in Arabic and Russian: say 'The sun yellows', 'The grass greens', 'Sugar sweetens' and 'Bears fur'. Expectations concerning 'The sun yellows' may become like those of 'The droplet glistens', 'The diamond sparkles', 'The star twinkles', etc.

Imagine this verbal idiom to be the only way of conveying such information. This is like Wittgenstein's conjecture: 'It would be possible to imagine people who, as it were, thought much more definitely than we, and used different words where we used only one.'[1] In such a language could it be stated as a fact that the sun *is* yellow, the grass *is* green, sugar *is* sweet? (Would this be like an observer's inability to see the antelope, having never before seen anything but birds?)

That it is yellow is a passive thing to say about the sun, as if its colour were yellow as its shape is round and its distance great. Yellow inheres in the sun, as in a buttercup. 'The sun yellows', however, describes what the sun does. As its surface burns, so it yellows. Now the grass would green; it would send forth, radiate greenness—like X-ray flourescence. Crossing a lawn would be wading through a pool of green light. Colleges would no longer be cold, lifeless stone; now they would emit greyness, disperse it into the courts. As a matter of optics this is rather like what does happen; the change of idiom is not utterly fanciful.

Grouping 'The sun yellows' with 'The bird flies' and 'The bear climbs' might incline us to view the dawn as a yellow surge over the horizon, a flood of colour enveloping the earth around us. Every student of optics at some time feels this shift in his concepts. This is not a case of 'say it how you please, it all means the same'. Kant remarked: '...they [who philosophize in Latin] have only two words in this connexion, while we [Germans] have three, hence they lack a concept we possess.'[2] This is not merely to speak differently and to think in the same way. Discursive thought and speech have the same logic. How could the two differ?[3] Speaking with colour-words as verbs just is to think of colours as activities and of things as colouring agents.

What if information about colours were expressed adverbially? We would then say 'The sun glows yellowly', 'The grass glitters greenly', 'The chapel twinkles greyly'. If everyone spoke thus how could one insist on its being a fact that the sun is yellow, that grass is green, or that the chapel is grey? Could such 'facts' be articulated at all?

It may be objected, 'However we speak, one could always see the difference between a bear climbing and grass greening. Language could not blind us to differences between the way the sun brings its colour to us and the way the waiter brings tea to us.' This is not obvious; it may not even be true. If we had no possible way of communicating about differences between bears climbing and grass greening, what would be the force of insisting that there were such differences?

'Indicating by means of language that some things cannot be thought is a suspect procedure. These differences in idiom are parasitic on differences existing in our present language. Anything one can indicate in our language *is* something that can be indicated in our language. The point of these linguistic experiments on the last five pages therefore dissolves.'[1] This objection has force. But have these experiments not at least suggested that people using different languages might have difficulty in apprehending the same facts? They are admittedly inconclusive. As a solvent of powerful presuppositions, however, they may do some good, and may even lubricate later arguments. Thus if, when confronted with fig. 3 of the last chapter, you see the bear while I see a tree from which limbs have been sawn, we have not made the same observation. If for you the fact is that the bear is brown, while for me it is that the bear browns, then we may not be aware of the same facts. Drawings of our data may be identical, yet our data differ. We begin from evidence which, though congruent, is disparate. This is the aspect of physics we must explore further, even if at times we feel like the eye trying to see the limits of its own vision.

What of 'primary' qualities? 'The sun is round' states a fact. So too 'St John's College hall is rectangular'; 'sugar lumps are cubes'.

Try 'the sun rounds', 'St John's hall rectangulates', 'sugar cubes'. Activity is suggested here. Would one who saw the round sun see the sun rounding? The college hall *is* rectangular. Would this fact be apprehended by a man for whom the hall rectangulates —holding itself in a rectangular form against gravity, wind, cold and damp? Perhaps the man for whom the sun rounds would see the sun incessantly arranging itself as a sphere. If he can say only 'The sun rounds', how else can he see it? Could one speak of candles being bright and yet see the candles *as* brightly?

If a conceptual distinction is to be made between a bear's activities when climbing and its activities when 'furring', the machinery for making it ought to show itself in language. If a distinction cannot be made in language it cannot be made conceptually.

Thus the logical structure of our conceptions of accelerating bodies is identical with the structure underlying 1–13 below:

1.	$a=(v-v_0)/t$	*definition*
2.	$v=v_0+at$	from 1
3.	$\bar{v}=s/t$ *or* $s=\bar{v}t$	from definition of average velocity
4.	$\bar{v}=(v_0+v)/2$	uniform change in velocity
5.	$\bar{v}=\frac{1}{2}(v_0+v_0+at)=v_0+\frac{1}{2}at$	via 2
6.	$\therefore s=\bar{v}t=(v_0+\frac{1}{2}at)\,t$	from 5
7.	or $s=v_0t+\frac{1}{2}at^2$	from 2, 6
8.	$t=(v-v_0)/a$	from 2
9.	$s=t(v_0+\frac{1}{2}at)$	from 7
10.	substituting $s=(v-v_0/a)\{v_0+a(v-v_0)/2a\}$	
11.	$s=(v-v_0/a)(v_0+v/2)$	
	and	
12.	$2as=v^2-v_0^2$	
13.	or $v^2=v_0^2+2as.$	

These are our concepts of accelerating bodies; thus is their behaviour described. Any distinctions to be made in our mechanical ideas must be made in this notation. How could an essential change be made here which left our conceptions of acceleration unaltered? This 'locking' of concept and language is fundamental in all physics.

Imagine people who count 'one, two, three, few, many'. Should we tell them that whether or not they can say so (think so, perceive so), it is a fact that St John's Tower has four spires? Perhaps for them that it has four spires is an inexpressible fact? Is this different from saying that, for them, it is no fact at all? 'This', it may be countered, 'reveals the paucity of their language. It is a fact that the Chapel Tower has four spires, because—even if everyone spoke and thought "one, two, three, few, many"—*were* there a language which could state what obtains in Cambridge architecture it would include "The Tower of St John's College Chapel has four spires".'

This objection, however, cuts into every language. Perhaps myriad things about St John's College elude English, and perhaps aspects of acceleration elude the algebra above? This is not to say that there are aspects of college chapels, acceleration, or anything else which *do* elude description; it is just not impossible that there might be. After all, we have the concept of things not being expressible in certain languages. Indeed we might have to come to think about the world differently, see different aspects of it, know 'facts' about it not even formulable now. Giving sense to this speculation is difficult; one cannot say the unsayable or express the inexpressible. One can only string together examples which may lend the suggestion plausibility.

This is not the hackneyed conjecture that the world might have been different. Given the *same* world, it might have been construed differently. We might have spoken of it, thought of it, perceived it differently. Perhaps facts are somehow moulded by the logical forms of the fact-stating language. Perhaps these provide a 'mould' in terms of which the world coagulates for us in definite ways.

We are familiar with acceleration as set out in the notation above; it is a modern analogue of Galileo's later arguments. But could Galileo's predecessors have had these ideas? Without the notation in which to express them? This is not a logical 'could': the argument is simply that the formation of a concept x in a language not rich enough to express x (or in a language which explicitly rules out the expression of x), is always very difficult. The conception of x without a notation in which x is expressible need not constitute a logical impossibility.[1] For example, in 1638 Galileo formed the concept of constant acceleration (which we express as d^2s/dt^2) while using a geometrical notation. Again, though Newton's fluxions symbolize ds/dt by '$\dot{v}$' only, he could move from differentiation to integration, something lesser men could accomplish only in Leibniz's notation. So it may not be impossible to think of x without a language in which x is expressible. Yet this is at least a practical limitation of the severest kind; and 'practical' does not mean 'conceptually unimportant'. This is not just a psychological point. It is of importance for understanding the way a natural philosopher thinks, and also for appreciating the nature of contemporary physical theory (within which what is difficult to conceive is sometimes formalized into what is impossible to conceive).

Both Galileo and his formidable opponent, Descartes, made the same mistake in a certain calculation, despite their disagreement on every other point. What they did not disagree about was their physical 'language'. This is worth consideration, for it reveals how intimately connected are physical concepts and the formalisms which express them.[2]

B

In a letter to Paolo Sarpi (1604) Galileo formulated a law of falling bodies. This was actually erroneous. In 1619 Beeckman sought Descartes' help with the same problem. Descartes' solution was similarly erroneous. Beeckman made yet a further error in interpreting Descartes' answer; and this inadvertently gave him the correct solution.

Galileo and Descartes made the same mistake in formulating this law. Why?

Galileo's letter argues thus:

> Considering the problems of motion—in order that I might finally show the phenomena I have observed to lead to an absolutely indubitable principle which I could then pose as an axiom—I came upon a proposition which seemed sufficiently natural and evident. If this proposition is postulated I can show everything else to follow; notably that the spaces traversed by a freely falling body are proportional to the square of the times, and that (in consequence), the spaces traversed in equal times are to each other as the series of odd numbers beginning with the number one, and other things as well. And the principle is this: that *the velocity of a freely falling body increases in proportion to the distance it has fallen from its starting point*; thus, for example, if a heavy body falls from *A* along the line *ABCD*, I regard the degree of velocity which it has at *C* to be related to the degree of velocity it has at *B* as the distance *CA* is related to the distance *BA*. Thus, in consequence, at *D* it has a degree of velocity greater than at *C* in just the measure that the distance *DA* is greater than the distance *CA*.[1]

This reasoning is typically Galilean. He seeks not a descriptive formula; nor does he seek to predict observations of freely falling bodies. He already has a formula.[2] He knows that the spaces traversed in equal times are to each other as the series of odd numbers. He knows that the distance fallen by the body is proportional to the square of the times. He seeks more: an *explanation* of these data. They must be intelligibly systematized. Galileo must reason back from these to a more fundamental principle, from whose assumption these 'accidents' will follow. He has no confidence in observations which cannot be explained theoretically.

Galileo was not seeking the cause of the acceleration; that was Descartes' programme. Galileo wished only to understand. His law of constant acceleration (1632) is not a causal law.[3]

Galileo's error consists in this: the principle he adopts as evident and natural—that the velocities of a freely falling body are proportional to the distances traversed—could never lead to the law of falling bodies as he formulated it. It leads to an entirely different law, expressible only as an exponential function. Galileo could never have managed such a formulation with the mathematics at his disposal.

The correct statement of the principle is: *the velocities of a freely falling body are proportional to the times—not the distances*. Leonardo da Vinci knew of this. Yet it took Galileo a long time to discover his own mistake. Why? What was seductive about this early version? A glance at the conceptual background may help.

Tartaglia realized the importance of the point from which a falling body departs. Against Aristotle he argued that the farther a body falls the more its velocity increases.[1] J. B. Benedetti added that this resulted from the body's being its own '*causam moventem, id est propensionem eundi ad locum ei a natura assignatum*'. He continued:

> The increase in force is always proportional to the increase in the distance traversed naturally, the body continually receiving new *impetus*; since it contains within itself the cause of its movement, which is the inclination to return to its natural place out of which it finds itself as a result of violence.[2]

Benedetti showed his respect for Aristotle here. He showed some confusion also; he sought to blend the physics of the schoolmen with that of the Parisian impetus theorists. Such was the obscurity into which Galileo stepped.

One feature of Benedetti's thought, however, is important: like all impetus theorists he regarded movement as the effect of a force contained in the moving body. Therefore he could detach the idea of a body's motion from that of the point towards which the body proceeded, the *terminus ad quem*. This allowed him to treat the motion in isolation from the rest of the universe. The space in which bodies fall was, for Benedetti, not essentially a physical, but rather a geometrical space; it was a space of co-ordinates, not of mutually attracting bodies. Motion in a void was thus quite conceivable for him. (The space was not *entirely* geometrical of course. There were two privileged directions: up and down.)

Impetus was a force impressed on the body by an external mover. This force could persist, so accounting for continued motion when bodies fell, or were thrown. Now what would happen if new impetus were impressed upon a body moving under its own impetus? From consideration of this query there followed a tolerable explanation of accelerated motion.[1] Many seventeenth-century physicists held this view—Piccolomini and Scaliger among others.[2] They treated the impetus as an efficient cause whose effect was the body's movement, the impetus being dissipated in this movement. Acceleration thus required the intervention of a new push or pull whilst the anterior impetus was still effective.

The theory was inconsistent. In the first moment after the removal of a body's support the impetus was held to result from an external cause—gravitational attraction. During later instants, however, the impetus itself was taken to cause movement.

Temporal and spatial aspects of impetus were clearly distinguished by da Vinci: 'The freely falling body acquires with each degree of time a degree of motion, and with each degree of motion a degree of velocity.'[3]

Why did Leonardo, Benedetti, and Varron assert the proportionality of the falling body's velocities to the spaces traversed, and not to the times? Doubtless they regarded these as equivalent. In another connection Duhem argued that to transform the law: *the distance fallen by an unsupported body is proportional to the square of the time fallen*, into the law: *the velocity of a falling body is proportional to the time of its fall*, one requires the idea of an instantaneous velocity, that is, the notation of the fluxion or the derivative.[4] Similarly here, to have detected the non-equivalence of points of time with points of space Leonardo, in the above quotation, would have needed the concepts of the integral calculus; so there is a sense in which the historical Leonardo could not have appreciated the facts of dynamics. When these predecessors of Galileo think of prolonged movement, they slide from temporal notions to those of space, from a movement's duration to its trajectory. 'Il est plus facile—et plus naturel—de voir, c'est-à-dire, d'imaginer, dans l'espace, que de penser dans le temps.'[5] Given the alternatives (1) velocities are proportional to the times, and (2) velocities are proportional to the spaces traversed, Leonardo, Benedetti, Varron, Galileo (1604) and Descartes all chose (2). The thinking of scientists in this period ran along geometrical rails; it was constituted of ideas of spatial relations. A 'time co-ordinate' would have had little significance for these natural philosophers, as little as would a 'fragrance' or a 'beauty' co-ordinate. Galileo broke through these conceptual limitations, but not until much later than 1604.

Benedetti represents the trajectory of a falling body as a vertical line *A—B*. Must not one admit that the body falls more quickly the farther it departs from *A*? and its velocity at *B* will increase if *A* is raised. This was known to sixteenth-century engineers: pile-drivers, as their heights were increased, drove a given weight farther into the ground. Is it not natural to consider this velocity at *B* a function of the space *AB*? Benedetti's was a most natural way to begin. At distances of less than 50 ft., differences between the elapsed times are not as likely to capture one's attention as are differences between distances, or between velocities. Benedetti's diagram gives time no prominence. How reasonable, then, to regard the fundamental relationship as being between the variation in height and the increase of velocity. The latter may even come to be regarded as a strict function of height. The final velocity indeed *does* depend on the initial height. To determine the exact dependence, however, requires mathematical techniques more subtle than anything known in 1604. The crude functions settled upon then would be more like: 'Double the initial height of a falling body and its final velocity will double.' This is erroneous. Yet given these premises, the suggestion that a body's velocities depend not on the distances fallen but on the times fallen would have appeared uselessly complicated. Time seemed but a trivial function of velocity.[1]

A

B

Indirectly, this attitude drove Galileo from the impetus theory. A minor merit of the latter is that it is time-dependent. The successive actions of impetus occur primarily in time; they are in space in only an inessential way. Ignore the causal aspect of free fall, however, and our ideas veer away from the temporal frame in which these causes successively act. Thinking becomes fixed to the spatial frame in which the 'resultant' motion is manifested. Galileo wished to geometrize motion, so he ignored this feature of the impetus theory, and hence the impetus theory as a whole. The impetus theorist sought to explain why motion continued, what caused it to go on; but that motion naturally continued was an unquestionable and unanalysable fact for Galileo. His thinking became orientated in a spatial framework, not in a causal, time-dependent one. Why does motion cease? That was Galileo's problem. Even as a young man at Pisa he realized that no geometrical system of dynamics could be founded on the intractable concept of impetus.

Galileo tried first to mathematicize Aristotelian physics. He failed. He then tried to rebuild physics on the idea of impetus. Again he failed. How could anyone determine the properties of so diaphanous, elusive and confused an idea? For this concept of an internal motive power he tentatively substituted the notion of repeated external attractions, or shocks, producing each new effect.

This march of Galileo's reasoning out of the wilderness in which Benedetti, Cardan and Tartaglia had floundered is instructive. For him their work rested on a contradiction: they sought a constant cause (impetus, gravity, etc.) to produce a variable effect (the velocities manifested by the falling body). His break with tradition was essential for his later retroductions[1] from phenomena, *li accidenti*, to principles which could explain those phenomena. It parallels Kepler's break with the tradition which required planets to move in perfect circles or in paths which were combinations of circular motions.[2]

By allowing an increase in acceleration when a body was under the action of a constant cause, impetus theorists admitted creation *ex nihilo*; and even in a consistent theory this could not be represented geometrically. (Thus did Benedetti rule out the influence of other bodies on the moving body. He chose to deal with motion in geometrical, not physical space, even though he would not completely abandon impetus.)

Galileo continued to use the term 'impetus'.[3] Now, however, it became the effect of motion, not some obscure cause. At Pisa (when studying the circular motion of the planets) he was already following similar conjectures by Piccolomini and Buridan. He saw that its impetus was not what moved Mars; this was but an aspect of Mars' motion being conserved. Velocity became a defining property of moving bodies; it was not sustained by something more fundamental. At Padua Galileo developed the idea of a *moment*—weight times velocity—a turning point in his thought.[4] Movement is simply substituted for impetus. It has become now a given fact, not a perplexing *explicandum*.

In 1604 Galileo knew that the spaces traversed by a falling body in equal times are to each other as the series of odd numbers, and he knew too that the distance fallen is proportional to the square of the times. He possessed the principle of the conservation of movement and velocity. All further attempts to find a causal account of free fall he renounced; the goal now was only an *explicans* which would unify and systematize these things he already knew.

The decision to treat motion geometrically removed its causes from further consideration and so took attention from time. Galileo wrote:

> I assume (and perhaps I can prove it) that a freely falling body unceasingly augments its velocity as its distance from its starting point increases; thus, for example, if the body leaves point *A* and falls along the line *AB*, I take it that the degree of velocity at point *D* will be greater than the degree of velocity at *C*, and that the distance *DA* is greater than the distance *CA*; thus the degree of velocity in *C* will be to the degree in *D* as *CA* is to *DA*, and thus in each point of the line *AB* the body will possess a degree of velocity proportional to the distance of that point from *A*. This principle seems very natural to me, and it comports very well with what we experience with machines and with instruments which act by shock, where the shock is made effectively greater as the height from which the body falls is increased. And if that principle is admitted, I can demonstrate all the rest. 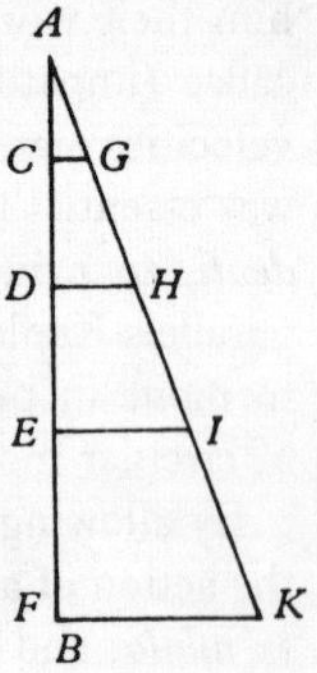
>
> Let the line *AK* make any angle with *AF*—let the parallels *CG*, *DH*, *EI*, *FK* be drawn from the points *C*, *D*, *E*, *F*. And since the lines *FK*, *EI*, *DH*, *CG* are to each other as are the lines *FA*, *EA*, *DA* and *CA*, it follows that the velocity in the points *F*, *E*, *D*, *C* are as the lines *FK*, *EI*, *DH*, *CG*. Thus the degree of velocity increases at every point in the line *AF* according to the increase of the parallels drawn from these same points. Hence, the velocity with which the body travels from *A* to *D* is composed of all the degrees of velocity which it acquires in all the points of the line *AD*—and since the velocity with which it traverses the line *AC* is composed of all the degrees of velocity which the body acquires in all points of the line *AC*, it follows that the velocity with which it traverses the line *AD* is to the velocity with which it traverses the line *AC* in the same proportion as all the parallel lines taken from all the points of the line *AD* to *AH* are to all the parallels taken from the line *AC* to the line *AG*;

and that proportion is just that of the triangle *ADH* to the triangle *ACG* —that is, as the square of *AD* is to the square of *AC*; thus the velocity with which the line *AD* has been traversed, is the velocity with which *AC* has been traversed in proportion double that of *DA* to *CA*. And since the relation of velocity to velocity is the inverse of the relation of time to time (because increase of velocity is the same thing as a decrease of time), it follows that the duration of movement following *AD* is to the duration of movement following *AC* inversely as the square of the distance *AD* to that of *AC*. The distances from the starting point are as the squares of the times, and consequently, the spaces traversed in equal times are to each other as the series of odd numbers beginning with one; this corresponds to what I have always said, and [also corresponds with] all observations. And therefore all truth accords with it. And if these things are true, I can demonstrate that velocity in a violent movement decreases in the same proportion as in a natural movement it increases, along the same straight line.[1]

This is plausible. The total velocity of the body is the sum of the instantaneous velocities acquired at each point of space fallen; it is also the sum of the instantaneous velocities acquired at each moment in the fall. But these 'sums' are not strictly comparable. The sequence of moments cannot be charted (on triangles) against the sequence of spaces 'which are to each other as the series of even numbers beginning with 1', because the 'sums' of the velocities increase as a linear function of the space traversed. Triangular representation allows one only to set out this uniform increase in relation to time.[2]

The notation blinds Galileo to all this. He transfers to space what belongs to time. Having made his first line stand for what he sees (namely, the fall of a body from *A* to *B*) instead of what he endures, it was inevitable that he should plot velocities against distances, not times. It is about velocity that he seeks information and the fall from *A* to *B* that he observes. When these are fitted together simply as the legs of a triangle there are no *logischen räume* for a time parameter.

Galileo ultimately noted his error; Descartes never did. In 1619 Beeckman asked Descartes why bodies fell.[3] The query is itself significant: causal ideas of Gilbert and Kepler are mixed in here. Bodies fall because the earth attracts them. Why do they accelerate? Beeckman's tentative answer was: because at each instant the earth attracts them anew, it confers on them an additional degree of motion. In 1613 Beeckman had formulated his own law of the conservation of motion:

Bodies once set in motion do not come to rest unless externally impeded. All things set in motion are like this. The weaker the impediment the longer the bodies remain in motion. If a body is projected upwards from a station which is itself in circular motion, so far as the upward motion is concerned the body does not come to rest before returning to earth; and when it does come to rest it does not do so because of evenly distributed impediments, but because of unevenly distributed impediments, since successive sectors of the earth come in contact with the moving body.[1]

Beeckman understood the physics of the problem. But he could not cope with the mathematics. Thus he asked Descartes:

Granting my principles—that a moving body will move eternally in a void—and supposing there to be a void between the earth and the stone which falls, can one determine the distance the stone will fall in one hour if one knows how far it falls in two hours?[2]

Beeckman interpreted Descartes' answer as follows:

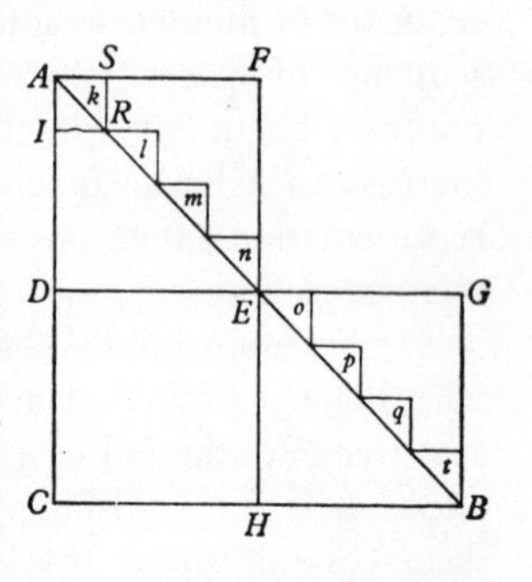

But, since these moments are indivisible the space traversed in an hour by the falling body will be *ADE*. The space traversed in two hours doubles the proportion of time, i.e. *ADE* to *ACB*, which is the double proportion of *AD* to *AC*. Let the moment of space traversed by the body in an hour be of any magnitude, e.g. *ADEF*. In two hours it will complete three such moments i.e. *AFEGBHCD*. But *AFED* is composed of *ADE* and *AFE*; and *AFEGBHCD* is composed of *ACB* with *AFE* and *EGB*, i.e. with the double of *AFE*.

Thus, if the moment is *AIRS* the proportion of space to space will be *ADE* with *klmn*, to *ACB* with *klmnopqt*, i.e. the double of *klmn*. But *klmn* is much smaller than *AFE*. Since therefore the proportion of space traversed consists in the proportion of triangle to triangle, equal magnitudes being added to each term of the proportion, and since these equal added magnitudes decrease proportionately to the decrease in the moments of space, it follows that these added magnitudes will be reduced to zero. Such is the moment of space traversed by the body. It remains then that the space traversed by the body in one hour is to the space traversed by it in two hours as the triangle *ADE* is to the triangle *ACB*.

This was shown by M. Perron, when I gave him the opportunity by asking him whether it could be known how much space is traversed in an hour, if the space traversed in two hours is known, granting my principle that what is once set in motion (*in vacuo*) remains in motion eternally, and supposing that space between the earth and a falling stone to be a vacuum. If then it is shown experimentally that a body falls 1000 feet in two hours, the triangle *ABC* will contain 1000 feet. Its root is 100 for the line *AC* which corresponds to two hours. Bisection at *D* gives *AD*, i.e. one hour. To the double proportion *AC* to *AD*, i.e. 4:1, corresponds 1000 to 250, i.e. *ACB* to *ADE*.[1]

This is both elegant and correct; the spaces traversed are proportional to the squares of the times. But this was not actually Descartes' solution.

A few days ago I met an ingenious man who put to me the following question: A stone, he said, falls from *A* to *B* in one hour; the attraction of the earth is exerted perpetually with the same force, and loses nothing of the speed imparted to it by the previous attraction. Now, what moves in space moves, he maintained, perpetually. The question is, in how much time will it traverse this space.[2]

Descartes, unlike Beeckman, regarded all this as mere hypothesis; nor did he even understand the hypothesis clearly. Descartes continues:

I have solved the problem. In the right angled isosceles triangle *ABC* represents the space (the movement); the inequality of the space from the point *A* to the base *BC* represents the inequality of movement. In consequence the time required to traverse *AD* is represented by *ADE*; and the time required to traverse *DB* by *DEBC*: it should be noted here that the smaller space represents the slower motion, but *ADE* is the third part of *DEBC*: hence *AD* will be traversed three times more slowly than *DB*. The question could be posed somewhat differently: supposing the force of attraction of the earth to be equal to what it was initially: and a new force introduced while the preceding one remains. In that case the solution is by a pyramid.[3]

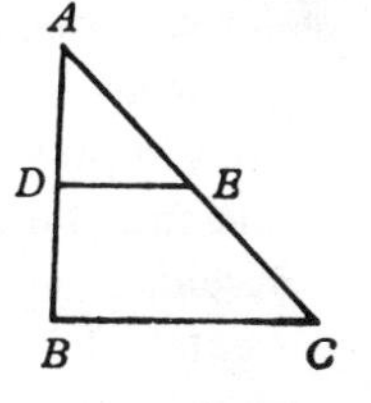

The point of the problem of free fall eludes Descartes. Beeckman's solution does not deter him. *Il ne laissait pas volontiers sa lumière sous le boisseau.* He proposes another possible case, one in which the attractive force grows from moment to moment. In the second moment of its fall a body is attracted with twice the force of the first moment, in the third moment with a triple force. In this solution the velocities increase in a cube-like way and not as squares.[1]

Descartes never asks about the physical possibility of this hypothesis of a growing force. It is a case of geometry—one more mathematical possibility. He is pleased to have found some relation between two sets of variables. Here is a pure geometer considering a problem of physical space and motion, and this is why he does not grasp Beeckman's principles, nor see the problem as does his correspondent.

The line *ADB*—which for Beeckman (as for the later Galileo) represented the square of the times[2]—is for Descartes, as for the Galileo of 1604, the trajectory of the unsupported body. This transforms Beeckman's problem: *ADB* is traversed with a velocity 'uniformly variable'. Descartes' problem is thus to find the velocity at each point in the trajectory.

For Beeckman *ADE* and *ABC* represented the spaces traversed, but for Descartes they are the sums of the velocities. Thus he concludes that the space *DB* was passed over three times more quickly than was *AD*. Time at last comes into his thinking here, but too late. The geometrization has, as with the young Galileo, with Benedetti and with Varron, already completely spatialized the problem: it has pushed time out. By understressing the physics of the problem (as this notation encourages one to do), Descartes joined the others in treating uniformly accelerated motion as that in which velocity increases in proportion to the distances traversed, and not in proportion to the times.

The time factor *could* receive due weight in this geometrical representation—the successes of Beeckman and the later Galileo prove this. But it is understandable why this factor should so long have been overlooked: thinking new thoughts in a conceptual framework not designed to express them requires unprecedented physical insights. In the history of physics few could sense the importance of things not yet expressible in current idioms. The task of the few has been to find means of saying what is for others unsayable. There was no derivative with which Galileo could attack the problem of constant acceleration, yet ultimately he fought through to what is essentially just this unifying idea. Newton had to build the whole theory of fluxions to speak the truth fully. Physics today is a mountain of mathematical formalisms; some of these express, explain and unify our observations.

Descartes' formalism was mathematically elegant, but it failed on the physical side:

In the question posed—where it is supposed that a new force is added at each instant to the one with which the heavy body tends towards the earth—this force increases in the same way as the transversals *DE*, *FG*, *HI*, and infinite other transversals which can be supposed to run between those specified. To show this I will suppose the square *ALDE* as the first minimum or point of movement, caused by the first imaginable attractive force of the earth. For the second minimum of motion

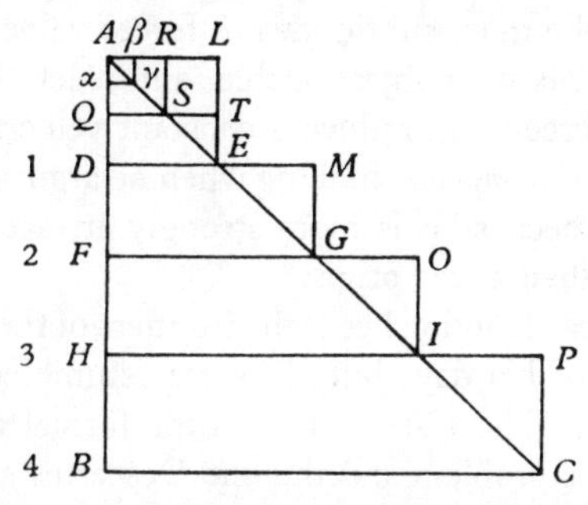

we shall have the double, namely *DMGF*: since the first force, which was the first minimum, remains, and a new and equal one is added to it. Similarly in the third minimum of motion there will be three forces, viz. those of the first, the second and the third minimum of time, etc. Now this number is triangular (as I shall perhaps explain elsewhere at greater length), and it will appear as representing the figure of the triangle *ABC*. But, you will say, there are protuberant parts *ALE*, *EMG*, *GOI*, etc., which come out of the triangle. Hence it will not be possible for the figure of the triangle to represent the progression in question. To which I reply that these protuberances are to be attributed to the fact that we have extended these minima which should be thought of as indivisible and not composed of any parts. This can be shown as follows. Let the minimum *AD* be divided into two equal parts at *Q*; then *ARSQ* will be the first minimum of motion, and *QTED* the second minimum of motion, in which case there will be two minima of forces. Similarly we shall divide *DF*, *FH*, etc. We shall then have the protuberant parts *ARS*, *STE*, etc. They are clearly smaller than the protuberance *ALE*. Let us go further. If, for a minimum, a smaller minimum is supposed, such as $A\alpha$, the protuberances will be even smaller, viz. $\alpha\beta\gamma$, etc. If, finally, for this minimum I take the real minimum, i.e. the point, then the protuberances disappear, because they cannot be the entire point, but only half the minimum *ALDE*; thus, as is evident, half a point is nothing. From all of which it follows that if we imagined a stone drawn by the earth in the void from *A* to *B* by a force emanating from it eternally at an even rate, while what has been accumulated remains, the first movement in *A* would be to the last movement which is in *B*, as the point *A* is to *BC*. As for the half *GB*, it will be traversed by the stone three times faster than the other half *AG*, since the stone would be drawn by the earth with a force three times greater. For the space *FGBC* is thrice the space *AFG*, as is easily proved. And thus, the same applies proportionally to the other parts.[1]

Descartes replaces Beeckman's principle of the conservation of movement with the more subtle idea of force: velocity is proportional to force. This is unobjectionable, and it lets him conclude that a constant force will produce a constant velocity. However, he tumbles back into *impetus* thinking when he argues that a falling body accelerates because it is more strongly attracted at the terminus of its fall than at the origin.[2]

Beeckman's idea of motion lies at the frontiers of the mathematics and the physics of his day; but Descartes stumbles, despite his brilliant geometry. There seems to be little formal difference between Beeckman's problem and the one Descartes substitutes for it; but Descartes translates the results of his integration into spatial terms. He substitutes the trajectory, not the duration, as the argument of his function.

The fact that Beeckman never detected the differences between his solution (velocities are proportional to the times) and that of Descartes (velocities are proportional to the spaces traversed) poses one of the problems this essay purports to illuminate. Beeckman's reaction to geometry would naturally be that of a physicist. Descartes', on the other hand, would be that of a pure mathematician. Koyré describes this as a comedy of errors,[3] but it is really much more profound. It is of a piece with Tycho and Kepler seeing different things at dawn. It is like Kepler's use of the same ellipse first as a mathematical tool and later as a physical hypothesis: the conceptual difference between these was enormous.[4] It is similar to the juvenile Galileo (1604) and the mature Galileo (1638) observing the same body fall, first ignoring and later appreciating the temporal aspect of the phenomenon. It is like a Mach and a Hertz getting the same formal solution to some dynamical problem; but what different solutions they really are.[1] It is like Born and Schrödinger dealing with the symbol 'ψ' in a mathematically identical way, but differing profoundly over its interpretation. It is like Einstein, De Broglie, Bohm and Jeffreys on the one hand, and Heisenberg, Dirac, Pauli and Bethe on the other, all considering the uncertainty relations; the latter group asserting and the former denying its indispensability in modern particle theory.

Ten years after this important exchange with Beeckman, Descartes takes up the problem again in a letter to Mersenne:

> Le sieur Beecman vint icy samedy au soir et me presta le livre de Galilée; mais il l'a ramporté à Dort ce matin, en sorte que je ne l'ay eu entre les mains que 30 heures. Je n'ay pas laissé de le feuilletur tout entier et je trouve qu'il philosophe assés bien du mouvement, encore qu'il n'y ait que fort peu des choses qu'il en dit, que je trouve entièrement veritable; mais, ce que j'en ay pû remarquer, il manque plus en ce ou il suit les opinions desia receues, qu'en ce ou il s'en esloigne. Excepté toutefois en ce qu'il dit du flus et reflus, que je trouve qu'il tire un peu par les cheveus. Je l'avois aussy expliqué en mon Monde par le mouvement de la terre, mais en une façon toute différente de la siene.
>
> Je veus pourtant bien avouer que j'ai rencontré en son livre quelques —unes de mes pensées, comme entre autre deus que je pense vous avoir autrefois escrites. La première est que les espaces par où passent les cors pesans quand ilz descendent, sont les uns aux autres comme les quarrés des tems qu'ilz employent à descendre, c'est-à-dire que si une bale employe trois momens a descendre depois *A* jusques a *B*, elle n'en employera. Qu'un a le continuer de *B* jusques a *C*, etc., ce que je disois avec beaucoup de restrictions, car en effect il n'est jamais entiérement vray comme il pense de demonstrer....[2]

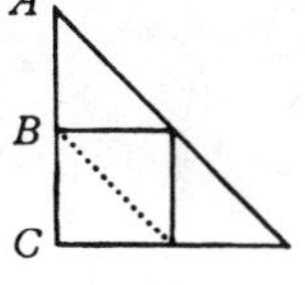

Descartes still regards the velocity of a falling body as a function of the space traversed. The solution of this physical problem was thus left to Galileo who, after 1604, began to 'see through' his notation.

CHAPTER III

CAUSALITY

> *We do not have a simple event* A *causally connected with a simple event* B, *but the whole background of the system in which the events occur is included in the concept, and is a vital part of it.*
>
> BRIDGMAN[1]

A

'For want of a nail a shoe was lost; for want of a horse a rider was lost; for want of a rider a battalion was lost; for want of a battalion a battle was lost; for want of a victory a kingdom was lost—all for want of a nail.' Here is a persistent view of causality. All of us hold it some of the time; some hold it all of the time. But it is inadequate for appreciating causal situations in physics.

The view can be put forward in several ways. One of these is the causal-chain figure. Consider Galileo with his inclined plane. The balls, terminating their descent and subsequent roll, collect in a loose formation at the far end of the floor. Down comes a brass sphere. It collides with another ball, which moves off with a predictable velocity. Again another ball is nudged: another, and another, always with a predictable velocity.

A better causal-chain account could not be found; but Russell comes close:

> Inferences from experiences to the physical world can...be justified by the assumption that there are causal chains, each member of which is a complex structure ordered by the spatio-temporal relation of compresence....[2]

> The chain of causation can be traced by the inquiring mind from any given point backward to the creation of the world.[3]

It is not language with which we are concerned. It is the concepts underlying this language.

Causal chains consist of links. They are discrete events, bound to neighbour-events very like themselves. 'All the members of such a chain are similar in structure...' (Russell).[4] Why did this ball move? That other ball struck it. Why did that ball strike the first? Because it was hit by another at a right angle. What made the third ball move? The brass sphere hit it after rolling down the plane. And so on. A variation is the genealogical-tree account. Broad appeals to this in his lively phrase 'causal ancestry'. Galilei senior was the father of Galileo and the son of Galileo's grandfather, *X*. *X*, in his turn, was the son of *W*, Galileo's great-grandfather, who was the son of *V*. Thus back to *A*, Adam. Bachelors break chains of succession, but in causal ancestry, every event has a cause and some effect: there are no bachelor-events. Note our temptation to think of nature as divisible into discrete happenings, each of which has one 'father' (cause) and one, or several 'sons' (effects).

This way of looking at the world leads to bewhiskered questions. *Y* caused *Z*, *X* caused *Y*, *W* caused *X*, *V* caused *W*. Thus back to *A*. What caused *A*? Dryden makes the standard answer pleasing to the ear:

> ...Some few, whose Lamp shone brighter, have been led
> From Cause to Cause, to Nature's secret head;
> And found that one first principle must be;
> But what or who, that Universal He;
> Whether some Soul encompassing this Ball,
> Unmade, unmov'd; yet making, moving All....[1]

Laplace claimed that, were he but supplied with an account of the state of the universe at one moment, plus a list of all the causal laws, he could predict and retrodict every other moment of the world's history—a dictum which is of a piece with the view to be examined.[2] However, for Laplace a causal chain was just a deductive chain. This complication merits special treatment.

Causes are related to effects as are the links of a chain, or the generations of a genealogical tree. It is all one plot with two themes, ancestry and progeny, like a novel by one of the Brontës. But this simplicity is unreal; and it springs from the same source as did the views of observation and facts examined earlier. Whatever else Galileo did, he did not dig up clues about the world in this simple fashion. Laboratory work seldom proceeds like the following-out of instructions on a treasure-map: 'ten steps north from the dead oak, four paces left, do this, now that, until at last the treasure, the cause.'[3] The tracts, treatises and texts of the last three hundred years of physics rarely contain the word *cause*, much less *causal chain*. In their prefaces and their *obiter dicta* physicists may get expansive;[1] nonetheless the concept is used infrequently in the actual practice of physics, and this fact is important.[2]

Why should this be? Because in so far as the chain analogy dominates a physicist's off-duty thinking about causation, he will find little in his work for which 'cause' seems the appropriate word. If he is free from thinking about causation in this way, other expressions may still seem more appropriate to his research. The elements of that research are less like the links of a chain and more like the legs of a table, or the hooks on a clothes pole. They are less like the successive generations of an old family and more like the administrative organisation of an old university.

Causal-chain accounts are just plausible when we deal with fortuitous occurrences, a series of striking accidents. Imagine Galileo at sunset packing his instruments. His telescope slips and begins to roll down the hill. The shrubs will stop it, he thinks, or the ravine. He sets off in pursuit. He does not notice a hole before him. He falls, and gets to his feet just as the telescope rolls past where the shrubs *had* been. (So that was what the gardener was up to.) And the ravine? Yes, it has been filled in too. Into the river goes the telescope. If only he had not been so clumsy; if only that hole had not been there, or at least had been seen; if only the shrubs had not been cut and the ravine had not been filled in. Galileo might even muse: for want of a ravine or a shrub there was nothing to stop the rolling telescope, for want of light I did not see the hole. So the instrument slipped into the river, all for want of a ravine or a shrub.

Or suppose that Galileo's carriage strikes a pedestrian in the darkened streets of Padua. The coroner might consider the circumstances: if only that banana skin had not been on the kerb; if only the driver had not been glancing back; if only the rivets in the brakeblocks had been secure. He too might set out his report: for want of a rivet a brakeblock was lost, for want of a brakeblock the distracted driver could not stop in time, for want of this control the carriage struck the Paduan who had slipped into the street because of the banana skin; this resulted in death—all for want of a rivet.

Furthermore, except in a context like the inclined plane experiment it would have been fortuitous that there was another ball *in situ* for the brass sphere to hit as it came off the plane, fortuitous that yet another ball was in the path of the ball set moving by the brass sphere....Do good billiards players achieve their results by accident? If they do not, causal-chain accounts of their performances are oversimplified. The 'for-want-of-a-nail' story is a chapter of accidents,[1] as is the 'for-want-of-a-ravine' and 'for-want-of-a-rivet' stories. An inquest of the street accident would consist in a coroner's recital of accidental happenings. Event *A* would not normally result in event *B*, but only in these exceptional circumstances; likewise for *B* and *C*, *C* and *D*, etc. The best materials for the causal-chain model are series of unparalleled events. But there are no inquest-laws of physics, for physics is not just a recording of dramatic accidents. Philosophers whose thinking about science is chained down to this notion of causation make laboratory research sound like an inquest.

B

Reference to one link of a chain *simpliciter* explains nothing about any other link—why, how, or from what it was made, etc. It does not even entail the existence of any other link. However, to know why the ball in Galileo's experiment was moved as it was, it is not enough to know that the brass ball was moving towards it just before a 'click' was heard. One must also know what are some of the properties of brass balls, and those made of other materials.[2] A familiarity with the dynamics of elastic bodies is involved too. Few would expect a head-on collision to result in both spheres moving off as one. All of us know enough dynamics to play golf, tennis, cricket, and to be able to make general comments about the spheres of Galileo's experiment (for example, we expect them to roll down the plane, not melt down it like hot wax, or transport themselves down like a water droplet on an oily slope).

To know why the kingdom was lost it is not enough to know that a battle was fought, that a battalion and a rider fared badly, that a horseshoe-nail was missing. It is also necessary to be familiar with the frictional properties of nails imbedded in cartilaginous substances, to know why horses are happier when shod, why dispatch carriers require horses, how helpless an isolated battalion can be, how much an army's fortunes can depend on one battalion, and the ways in which the security of kingdoms can depend on military success. To understand how the Paduan pedestrian came to grief, it is not enough to know that certain incidents were strung out in temporal order $t_0, t_1, t_2, t_3, \ldots$. One must know what usually happens when people step on banana skins, when drivers are distracted at dusk, when the rivets on brakeblocks are insecure.

The primary reason for referring to the cause of *x* is to explain *x*. There are as many causes of *x* as there are explanations of *x*. Consider how the cause of the death might have been set out by a physician as 'multiple haemorrhage', by a barrister as 'negligence on the part of the driver', by a carriage-builder as 'a defect in the brakeblock construction', by a civic planner as 'the presence of tall shrubbery at that turning'.[1] The chain analogy obscures this feature of causation. Examples adduced in favour of the analogy (billiard balls colliding and levers opening switches) are tacitly loaded with assumptions and theoretical presuppositions. Without these the examples would not be intelligible, much less support one view against another. Only its simplicity and familiarity makes this background knowledge fade before the spectacular linkage of the attention-getting events.

Nothing can be explained to us if we do not help. We have had an explanation of *x* only when we can set it into an interlocking pattern of concepts about other things, *y* and *z*. A *completely* novel explanation is a logical impossibility. It would be incomprehensible (just as a completely sense-datum visual experience is a patchwork of colours, wholly without consequences); it would be imponderable, like an inexpressible or unknowable fact.

The chain model encourages us to think that only normal vision is required to be able to see the brass sphere causing another ball to recoil away; apparently one has only to look and see the linkage between the missing nail and the collapsing kingdom, or the loose brakeblock rivet and the untidy Paduan street.[2] But in fact what we refer to as 'causes' are theory-loaded from beginning to end. They are not simple, tangible links in the chain of sense experience, but rather details in an intricate pattern of concepts. Seeing the cause of the movement of the stars, or the coolness of the night air, is less like seeing flashes and colours and more like seeing what time it is, or seeing what key a musical score is written in, or seeing whether a wound is infected, or seeing if the moon is craterous. Let us consider this further.

'The scar on his arm was caused by a wound he received when thrown from his carriage.' Here 'wound' is an explanatory word; 'scar' serves (here) not as an explanation, but as an *explicandum.* What we call 'wounds' and 'scars' are seldom strung on the same chain of discourse by a repetitive linkage: the situation is more complicated. What is the difference here between 'scar' and 'wound'? What is it for a man to have been wounded?

Is a wound just a more-than-superficial incision? Let us agree that it is. Minor scratches and nicks will not count as wounds. However, surgeons do considerably more than scratch and nick their subjects; yet it is not usual to speak of a surgeon as wounding his patients.[1] Does an operation on a fully anaesthetized patient count as wounding when the incision has been planned after consultation with other experts? No. Does the surgeon inflict a wound when he drops a scalpel, cutting a patient's arm? Perhaps. More must be known about the situation. Does the plantation owner wound the rubber tree when he carves the V-shaped trough deep into its bark? Again, perhaps. More needs to be learned about *Ficus elastica* before we can say. We can be certain of this, however: the Eskimo hacking blubber off a dead whale is not wounding the whale. Nor will throwing darts at a stuffed moose head upset the R.S.P.C.A. The carpenter does not wound the timber, however much he slices, gouges, and drills it. Only living things can be wounded; no incision in dead matter is a wound.[2] Nor would every deep incision in a living organism be a wound. An anaesthetized man undergoing appendectomy is not being wounded. A deep cut in a calloused foot is not a wound. Carving identification marks in the horns and hoofs of cattle is not wounding them.

A wound, therefore, is not *any* sort of deep incision: it is one which endangers the life, or impairs the functions of the wounded. That is why it cannot be said whether the ministrations of the plantation owner constitute wounding *Ficus elastica.* His slicing may impair the plants' functioning and endanger their lives; more information about the species is required before we can decide.

For the person who asks 'What caused that man's scar?', the scar is a visible datum: it can be seen. It is an *explicandum* about which he asks this and other questions. He could sketch the scar. But for that same man a sketch of the original wound may be nothing more than a picture of a deep incision. To see it as a wound is to identify it. It is to diagnose it as endangering life or impairing function. To see a wound at all requires knowing a modicum of pathology. In other contexts seeing scars requires knowing some dermatology and neurology. For instance, 'What caused the instrument-maker's retirement?'. 'His fingertips were scarred in a carriage-accident.' To see his scarred fingertips is to see why he could no longer build instruments. Merely to see his fingertips as rough and calloused is still to require information about the effects of such tissue on one's dexterity.

This feature of causation and explanation gets lost when concepts are forged in the causal-chain mould. The scar on a man's arm is explained by reference to the wound which caused it, because a wound is the sort of deep incision that would leave a scar like that. To hang the wound and the scar on the same causal line fails to mark how scars are explained by reference to wounds. 'Scar' and 'wound' are words on different theoretical levels.

Galileo often studied the moon. It is pitted with holes and discontinuities; but to say of these that they are craters—to say that the lunar surface is craterous—is to infuse theoretical astronomy into one's observations. Is a deep, natural valley a crater? Miners dig steeply and deeply, but is the result more than a hole? No; it is not a crater. An abandoned well is not a crater; nor is the vortex of a whirlpool. To speak of concavity as a crater is to commit oneself as to its origin, to say that its creation was quick, violent, explosive: artillery explosions leave craters, and so do falling meteors and volcanoes. Sketches of the moon's surface would just be sketches of a pitted, pock-marked sphere; but Galileo saw craters.

Again, a liquid is tasted and is declared to be bitter. That is all the tongue can tell, but we, perhaps, say that it is poison. To say of a liquid that it is poison is to diagnose it as capable of doing all the things poisons do. 'What caused the death?' We might answer 'poison'. When would we answer 'a bitter liquid', and leave it at that?

A wound is a cut which endangers life or impairs function. A scar is a dermal discontinuity which lessens sensitivity, and sometimes dexterity. Lunar craters are superficial pits resulting from explosion or impact. Poison is lethal. Words like 'wound', 'scar', 'crater' and 'poison', are often expressed with medical, biological, geological, and chemical overtones. Diagnoses, analyses, prognoses, are built into them. That is why in certain contexts they explain scars, clumsiness, rough surfaces and death; why it is natural to refer to the wound as the cause of the scar, to the scar as the cause of clumsiness, to the crater as the cause of uneven surface reflexion, to volcanoes or meteors as the cause of the crater, and to poison as the cause of death. Scars are what most wounds result in; hence it explains a man's scar to say that it was caused by a wound incurred in a carriage accident. (Most words which serve in this explanatory capacity are loaded in a similar way: for instance, particle, elastic, vector, acid copper, eclipse, light...there is no end.)

The terms of physics thus resemble 'pawn', 'rook', 'trump' and 'offside'—words which are meaningless except against a background of the games of chess, bridge and football. To one ignorant of what happens as a rule in bridge, 'finesse' will explain nothing. Even though nothing escapes his view while the finesse is made, he will not *see* the finesse being made. To one ignorant of what happens as a rule with chemical solutions, 'laevo-rotatory' will explain nothing, though his gaze be fixed on all the laboratory equipment when the chemist makes his announcement. Similarly 'wound' explains the man's scar only against the implicit background of theory brought out here. So too with 'crater' and 'poison'. The diagnostic and prognostic quality of these causal substantives reflects in the verbs with which they combine, verbs which are loaded in the same way: 'inhale', 'perforate', 'dissolve', 'charge', 'expand', 'stretch', etc.

Consider 'stretch'. We stretch rubber, elastic bands, springs, shrunken clothing, our arms and legs. Do we stretch butter from one corner of a scone to the other? Do we stretch seed from one corner of the garden to another? Does the gas escaping from the cooker stretch into the atmosphere? Does a cloud stretch when caught by a wind? Perhaps these are all cases of stretching; there are times when these might be natural ways of speaking. Still, there are differences between stretching rubber and springs on the one hand and stretching butter, sand and gas on the other. In the former cases, when we stop stretching, the body returns to its original shape; but this is not so with butter, sand or gas. The cloud does not snap back to its earlier shape when the wind dies.

Shrunken socks do not return to their diminutive shapes after stretching; but yet we cannot stretch socks as we might 'stretch' butter, or sand, or gas. Waterfalls are not stretching water. A spreading population is not stretching.

Though rarely explicit, the diagnosis built into 'stretch' will differ, then, according to whether it is rubber, sand or the truth that we are stretching. There are many theoretical backcloths against which 'stretch' can show up.

To improve a clock's action a clock-maker may stretch its mainspring. He may stretch the lubricant available for the clock, and he may stretch the truth when showing the clock to a buyer. Doubtless he knows what behaviour is appropriate in each case. He possesses knowledge of mainsprings, stress, strain, and elastic limit; of lubricants, viscosity and the oil requirements of this clock; and of people who inquire after 'olde Englishe clockes'.

The fact that *x* has been stretched can explain some event *y*. Or, *x*'s having been stretched can be the *cause* of *y*. This is saying more than that if we had opened our eyes and looked, a picture of *x*-being-stretched-and-causing-*y* would have registered in our visual space. These suggestions may be reinforced by some further remarks on *seeing that*.

C

One of Galileo's apprentices (Viviani or the young Toricelli) may see how gears, rachets and levers of a clock engage each other. He may portray this in a diagram; and in this the apprentice may excel his old master (whose vision was weak and failed late in life). But Viviani may not yet see that the force transmitted by the weighted drum is passed on to the driving gear, and thence through the gear-train, to 'escape' by measured degrees through the escapement—something which old Galileo is sure to see. The young apprentice may not appreciate the dynamics of the weight suspended from the driving drum, nor how these are related to the instrument's activity.[1] That most of us *do* appreciate this—*do* see the weight which pulls the string which turns the drum which drives the gear train—does not make such knowledge less essential to our comprehension of 'The driving weight is the cause of the escapement's action'. The statement is intelligible only in terms of knowing something about the properties of metals, the elements of mechanics, and the principles of horology.

Seeing what causes a clock's action requires more than normal vision, open eyes and a clock: we must learn what to look for. We do not recite the lessons of this training each time we see the cause of some event, but their content is indispensable in the search. The chain account obscures this by ignoring it: it treats the world as a simple Meccano construction where observers are cameras. But causes are no more visual data *simpliciter* than are facts. Nothing in sense-datum space could be labelled 'cause', or 'effect'.[1]

Yet, unmistakably, the old Galileo sees what causes the clock's action. '"I see", said the blind man, but he did not see at all.' A blind man cannot see how a timepiece is designed, or what distinguishes it from other clocks. Still, he may see that, if it is a clock at all, it will embody certain dynamical principles; and may explain the action to his young apprentice. The latter, however keen his vision, can describe only the perturbations of the clock; he cannot say what causes it to behave as it does. Galileo can say what causes it (and any other similarly constructed clock) to do what it does, because the blind Galileo has what his apprentice lacks—a knowledge of horological theory. Though the apprentice has what Galileo lacks, normal vision, he cannot detect the cause of the clock's motion.

Notice the dissimilarity between 'theory-loaded' nouns and verbs, without which no causal account could be given, and those of a phenomenal variety, such as 'solaroid disc', 'horizoid patch', 'from left to right', 'disappearing', 'bitter'. In a pure sense-datum language causal connexions could not be expressed. All words would be on the same logical level: no one of them would have explanatory power sufficient to serve in a causal account of neighbour-events. But it is here that the causal-chain should work best, for at the sense-datum level all events *are* like the links of a chain. They meet Russell's requirement by being similar in structure. Yet they elude the language of causality.[2] The chain analogy is appropriate only where genuine causal connexions cannot be expressed. How could explanations be advanced in a sense-datum language?

It is not that certain words are absolutely theory-loaded, whilst others are absolutely sense-datum words. Which are the data-words and which the theory-words is a contextual question. Galileo's scar may at some times be a datum requiring explanation, but at other times it may be part of the explanation of his retirement. 'Wound' helped to explain Galileo's scar, but it might also express a datum, something observed yet requiring explanation—as when a medical classroom is bedecked with pictures of several varieties of wound, all awaiting commentary by the Professor of Surgery. 'Red now', 'smooth', 'disappearing' are not once-for-all unladen with theory. Such language could function within sophisticated explanations, rather than as mere verbal records of immediate experience. 'Red now' in an astrophysical context (involving the Döppler effect) might explain celestial phenomena. 'Smooth' in a statistical context, 'disappearing' in a cathode-ray tube context (involving, say, Crooke's dark space), likewise contain volumes. We can infer an effect from some cause only when the 'cause-word' guarantees the inference; but which words are cause-words and which effect-words is for the context to determine.

Causal connexions are expressible only in languages that are many-levelled in explanatory power. This is why causal language is diagnostic and prognostic, and why the links-in-a-chain view is artificial. This is why within a context the cause-words are not 'parallel' to the effect-words, and why causes explain effects but not vice versa. For 'cause'-words are charged: they carry a conceptual pattern with them. But 'effect'-words, being, as it were, part of the charge, are less rich in theory, and hence less able to serve in explanations of causes. Galileo might explain the action of the clock hands by reference to the weight-driven main gear, the gear train, the escapement, the pendulum. (Note how 'theory-loaded' are each of these expressions; how extensive are their horological and dynamical implications.) He might say that the main gear, the train, escapement, and pendulum cause the motion of the hands. Explanation would not proceed in the opposite way: Galileo would not explain the action of these parts by describing the motion of the hands. Neither the system of dynamics nor any system of horology unfolds in that order.

When the apprentice says 'pendulum-escapement', he may mean little more than 'tick-tock, to-and-fro'. Much may follow from that, but not what follows for Galileo when these same words leave his lips. When the youngster says 'lightning and thunder', he probably means 'flash and rumble'. Again, a lot may follow, but what follows for him is different from what follows for the meteorologist—for whom 'lightning and thunder' probably means 'electrical discharge and aerial disturbance'. Ask the shepherd 'what caused that thundering noise?' and the response may be 'rain is on the way'. The meteorologist says: 'The noise originates near that cumulus cloud. In principle the cloud is an electrostatic generator. The ice crystals within it produce, by friction between themselves, electric charges, the separation of which leads to a concentration of positive charge in one region of the cloud and of negative charge in another. As charge separation proceeds, the field between these charged centres (or between one of them and the earth) grows. Finally, electrical breakdown of the air occurs; we see this as lightning. It leads to a partial vacuum in the atmosphere. Surrounding air rushes in. The result is a disturbance not unlike the breaking of a lamp bulb; we hear this as thunder.'

One might regard this as an example of causal-chain talk. Rather, it is a deductive chain. Each step in this account does follow the one before, but not like links in a chain or sheep over a log. This is not the single-file following of children's games but the following of entailment; it is details following a pattern, elements following a scheme. Much more than normal vision is involved in seeing a flash as lightning, and in hearing a rumble as thunder.

The 'wider' a word is theoretically, the more loaded it can be causally. The more widespread its net of effect-words, the more fertile its explanatory possibilities. ('...["The sky looks threatening": is this about the present or the future?] Both; not side-by-side, however, but about the one *via* the other.')[1]

Cause-words resemble game-jargon, as was noted earlier. 'Revoke', 'trump', 'finesse' belong to the parlance system of bridge. The entire conceptual pattern of the game is implicit in each term: you cannot grasp one of these ideas properly while remaining in the dark about the rest. So too 'bishop', 'rook', 'checkmate', 'gambit' interlock with each other and with all other expressions involved in playing, scoring and writing about chess.

Likewise with 'pressure', 'temperature', 'volume', 'conductor', 'insulator', 'charge' and 'discharge', 'wave-length', 'amplitude', 'frequency', 'elastic', 'stretch', 'stress' and 'strain' in physics; 'ingestion', 'digestion', 'assimilation', 'excretion' and 'respiration' in biology; 'wound', 'poison', 'threshold' in medicine; 'gear-train', 'escapement', 'pendulum' and 'balancer' in horology. To understand one of these ideas thoroughly is to understand the concept pattern of the discipline in which it figures. This helps to show how cause-words are theory-loaded in relation to their effect-words. It is something like the way in which 'trump', a bridge-loaded word, explains 'beat my ace' which is not bridge-loaded, but merely more thinly game-loaded. 'You beat my ace' might be said in many card games; 'you trumped me' will be heard only in bridge and whist. The more 'phenomenal' a word, the less 'theoretical' it is. We are more capable of understanding these low-level words independently of the language-system in which they figure. Children in the nursery, after learning a few object-names[1] do quite well with 'cold', 'hot', 'red'. The more their experiences vary, the greater the demands put on the language they are learning to use. When explanation, causation and theorizing have become their daily fare, each element of their speech will have worked into a comprehensive language-pattern, buttressed and supported in many ways by the other elements. Questions about the nature of causation are to a surprising degree questions about how certain descriptive expressions, in definite contexts, coupled together, complement and interlock with a pattern of other expressions.

D

Context has been stressed. The background information, the 'set' that makes an explanation stand out, derives as much from what is obvious in a situation as from discursive knowledge gained through training. If someone opened the door and shouted 'Fire!', you would not have to rummage through your memory before suitable action suggested itself. This is connected with the remark that a body of theory and information guarantees inferences from cause-*words* to effect-*words*. Cause-words, in appropriate contexts, unleash much more than an isolated word in an indefinite context. If, in the blank pages of a next year's diary we find the word 'fire' in the place reserved for St Valentine's day, no action would suggest itself. Consider another man shouting 'Fire!'; but now he is in uniform, hovering over a busy gun crew. Were we members of that crew, our response would be automatic. (One thing we should not do would be to scurry for shelter.) In other contexts 'fire' might herald a worker's dismissal, or the entrance of a Wagnerian soprano amid pyrotechnics. It can signal a phase in the making of pottery, describe how an actress reads her part, or designate some primitive rite.[1] From these utterances (in specific contexts) much can be inferred. 'Fire' has, in each situation, a propositional force; it is shorthand for complex statements whose nature is clear from the contexts of utterance.[2] We are not born able to recognize such contexts, any more than we are to see eclipses and escapements. For that we need education.

This is more familiar than it appears. After moving his stethescope over a patient the physician exclaims 'valvular lesion'; his nurse understands this as an intelligible assertion. The chemist who knowingly labels a flask of water 'inflammable' will be pressed for an explanation. Words like 'lesion' and 'inflammable' in these contexts, like 'pendulum', 'wound' and 'lightning' in others, do the service of complete propositions. And propositions are the stuff of inference.

In a similar way, though not in the same way, 'cause' words show their family connexions in the contexts of their employment. They draw explanatory force from conceptual patterns underlying the situations in which they are used, somewhat as 'fire' draws propositional force from contexts in which it might be uttered. 'It takes a particular context to make a certain action into an experiment....But if a sentence can strike me as like a painting in words, and the very individual word in the sentence as like a picture, then it is not such a marvel that a word uttered in isolation and without purpose can seem to carry a particular meaning in itself' (Wittgenstein).[3]

Further, if a man shouts 'Fire!', pointing to a blazing dynamite warehouse, and then adds 'Run for your life!', we might say 'Naturally—what else?'. Part of the force of 'Fire!' here is that he who hesitates is lost. The added 'Run for your life!' is compellingly obvious. Who could hear and understand such an alarm and fail to run? Effect-words dovetail with cause-words like this. That the clock-hands are moved by the weighted gear-train will seem obvious if we know what gear-trains and clock-hands are. One may even feel that, in a sound clock, the hands, being what they are, *must* be moved by the gear-train—for would there not be something unsound about the clock if they were not? (The weight actuates the gear-train, and ultimately the hands. This is what we mean by 'clock', 'gear-train', 'weight', and 'hands'.)

This is the whole story about necessary connexion. 'Effect' and 'cause', so far from naming links in a queue of events, gesture towards webs of criss-crossed theoretical notions, information, and patterns of experiment. In a context and by way of a theory, certain effect-words inevitably follow the utterance of certain cause-words: 'main-spring uncoils—hands move', 'lightning flashes—thunder rumbles', 'rain falls—wet pavement', 'summer—heat', 'fire—destruction'.

Causes certainly are connected with effects; but this is because our theories connect them, not because the world is held together by cosmic glue. The world *may* be glued together by imponderables, but that is irrelevant for understanding causal explanation.[1] The notions behind 'the cause *x*' and 'the effect *y*' are intelligible only against a pattern of theory, namely one which puts guarantees on inferences from *x* to *y*. Such guarantees distinguish truly causal sequences from mere coincidence. There is no connexion between the swings of Galileo's pendulum and the synchronous ringing of the distant church-bell. Nor is there a causal connexion between baby's first taste of banana and a simultaneous eclipse of the sun, though this may put the child off bananas. Similarly there is no causal relation between my winding the clock and then going to sleep, though no two events occur with more monotonous regularity. One could predict my going to sleep from watching me wind the clock, or retrodict my having wound the clock from observing me asleep. But this is risky, like amateur weather-forecasting or angling advice. No conceptual issue is raised by the failure of such a prediction or retrodiction. Our understanding of nature receives no jar from guessing wrong here on one occasion.

This shows what we expect of a causal law. These are not built up in the manner: $(A \text{ then } B)_1$, $(A \text{ then } B)_2$, $(A \text{ then } B)_3$, therefore all *A*'s are followed by *B*'s. This obscures the role of causal laws in our conceptions of a physical world. It is not merely that no exceptions have been found. We are to some extent conceptually unprepared for an exception: it would jar physics to its foundations; the pattern of our concepts would warp or crumble. This is not to say that exceptions do not occur, but only that when they do our concepts do warp and crumble.[2] It is all or nothing. The causal structure of the universe, if such a thing there be, cannot be grasped simply by counting off event-pairs, Noah-fashion, and then summarizing it all with an umbrella formula.[1]

The difference between generalizing the repeated occurrence of contiguous, propinquitous, asymmetric event-pairs and understanding the 'causal' structure of a natural phenomenon is like the difference between having a visual impression of a lunaroid patch and observing the moon. It is like the difference between contemplating a concavity on the lunar surface, and appreciating the fact that the moon is craterous.

Coincidental event-pairs are bound by no reputable theory, and we would feel little unsettlement if one occurred without the other. That happenings are often related as cause and effect need not mean that the universe is shackled with ineffable chains, but it does mean that experience and reflexion have given us good reason to expect a *Y* every time we confront an *X*. For *X* to be thought of as a cause of *Y* we must have good reasons for treating '*X*', not as a sensation word like 'flash', 'rumble', 'bright', 'solaroid', 'bitter' or 'red', but rather as a theory-loaded, explanatory term like 'wound', 'crater', 'stretch', 'pendulum', 'discharge' or 'elastic impact'.

This is obscured by the links-in-a-chain, ancestry-progeny view of cause and effect. How could such a view ever grip us? Why do we so often think of physical events as clicking off in single file?

E

The first scientific theories were those of astronomy and mechanics. These apply to animate and inanimate objects alike. Causal explanations were from the start expressed in terms of impact, attraction, momentum—in short, pushes and pulls. This led to the notion that all causes were impacts, attractions, pushes and pulls, and all effects the result of pushes and pulls. The conviction that sooner or later all science is mechanics dies hard: for three centuries science has been dominated by notions of inertia, impact and resultant velocities. This has affected our understanding of causation.

Nonetheless the terms of classical physics are as theory-loaded as can be. In appropriate contexts words like 'force', 'equilibrium', 'component', 'translatory', 'momentum', 'position', 'displacement', 'velocity', acceleration', contain volumes. Only a hasty view of the vocabulary of mechanics will support a chain conception of causation; classical physics is not like a series of links in its simplicity. It would not have required a Galileo and a Newton to create it were that true. Why should the many-levelled character of explanations in dynamics have been so long overlooked in favour of an attitude transparently inappropriate once pressed past metaphor? What is it about physics which is so receptive to the causal chain model?

Several things conspire to make the model attractive. The first is connected with the rise and influence of physics: not with Galileo's subject-matter, but with his method. The 'mathematical-deductive method' gave Galileo's insight the clarity and power required for putting science on a sure foundation. Chains of reasoning, deductive chains, played a spectacular role here. Physicists thought of God as a master-mathematician. They thought the Pythagoreans correct: the structure of the world was essentially numerical. Natural philosophy became a mathematical undertaking; Euclid was read as a preface to the Book of Genesis.[1] By the eighteenth century, after the successes of Galileo, Kepler and Newton, the universe was construed as an intricate geometric-arithmetic puzzle. Nor was this unreasonable. To have seen how mathematical configurations were related and interlocked was, for the physicist, to have learned about how events in nature were related and interlocked. Just as the premisses and conclusions of a deduction were connected by a series of formal steps, so the causes and effects in a natural phenomenon (e.g. the communication of momentum by impact) were connected by a series of events—links in a causal chain. The multiplication table had allowed Galileo and Kepler to find unexpected formal relations between parameters not linked naturally in any obvious way—e.g. $s = v_0 t + \frac{1}{2}at^2$. Similarly the theory of functions was beginning to supply physicists with a store of possible relations between variables which they might encounter in observation. The ramifications of this are only now being felt (see ch. VI).

Thus the rigour and inevitability which marks a formal proof became detectable in nature. Causes were three-dimensional premisses, effects were three-dimensional conclusions, and the two were linked by intermediate events, as necessarily as $(\sqrt{(25)} + 1)$ equals $+6$ or -4. Causal chains were three-dimensional deductive chains.

'A chain is as strong as its weakest link' hints at another feature of the model. The adage is true of ordinary chains; and deductions, too, are only as strong as their weakest steps; while physical experiments are, of course, only as reliable as the least reliable phases of their design and execution. This indicates how the chain figure appeals to those for whom physics is like the investigation of clockwork. To strengthen a chain we must detect and toughen its weaker links. When a wrist-watch misbehaves, removal of the cover will disclose those flaws which brought about the stoppage: a fouled hairspring or a dislodged jewel. Likewise with mathematics we turn our attention to the steps where breakdown is most likely. (We all know the feeling of 'Eureka!' that comes from snapping a whole deduction into focus by one alteration.) So too with experiment. A physicist examines his apparatus thoroughly, yet his readings show something amiss. From long experience he knows where a miscalculation or clumsy construction is most likely to be found. He does not re-examine the tubing, the stoppers, the power or the insulation, which have proved themselves reliable. But that new contact breaker with all its Victorian delicacies—perhaps he should take a second look there. Physicists learn to detect troublesome steps; they 'strengthen their weakest links'. Any analogy that calls attention to this self-corrective aspect of experiment is valuable.

What is overlooked, however, is that experiments are *designed* to be as chain-like as possible.[1] It takes long training to master this technique of design. To bring together a cluster of theoretical considerations in a single, tersely-expressed hypothesis; to torture it in an experiment, each phase of which keeps everything constant except one set of factors; to insure that when these vary to a certain degree they initiate another phase of the experiment, where again everything is constant save one factor; to have arranged that these in turn play roles in further phases of the demonstration—not just anyone, not every physicist, can juggle all this into an efficient laboratory operation. To characterize such an enterprise as 'this happens, then that, then those things take place, which results in...', is a bad caricature. The balance, timing, ingenuity and planning involved in a first-class experiment (like the determinations of *g*, *e*, *h* and *c*) can be represented in links-in-a-chain fashion only by a casual observer.[2]

The chain analogy will satisfy only those who have been taken in by the dramatic effects of experiments, and who see only their spectacular surfaces. The effects are *contrived* to rivet attention on some select sequence, out of a complex of possible sequences, which is pertinent to the experimenter's purposes. In the inclined-plane experiment there were myriad alternative sequences which might have captured attention—the effect of the rolling spheres on local air currents, for example, or the effect of the sphere's weight on the surface of the grooved plane, or the effect of the success of the experiment on Galileo's pulse-rate. Philosophers who dwell exclusively on the attention-getting events fail to note what is involved in directing attention in some desired manner. It cost Galileo thirty years of labour before he saw the conceptual structure of acceleration clearly enough to confirm his ideas by the inclined-plane experiment. Even before a chain account of the 'surface' of an experiment is possible, Nature must have been tampered with. In Nature, unlike the laboratory, physical conditions are rarely held constant whilst certain factors are allowed to vary for the benefit of a well-placed observer. Even the simple inclined-plane experiment rests on constancy factors not to be found in any cluster of events occurring naturally.[1]

Suppose that conditions in nature *were* held constant. The chain analogy would still be artificial, since it would not indicate how the explanation of events came about nor in what this explanation consisted. So that even were the subject-matter of physics like that of the coroner—a string of accidents against a fixed background—the chain analogy could not indicate what is important about causal talk, namely its explanatory capacity.

Another encouragement for causal-chain thinking came with the designed machine. Some mechanisms create their own laboratory conditions; they are indifferent to alterations in environment. Clocks, anemometers, windmills, gyro-compasses, thermostats, are made not to stop for thunderstorms, swarms of bees, and barking dogs. Under way, they go on *proprio motu*. Perhaps these led us to construe causal explanations as accounts of the perseverance of manufactured machines, the movements of stars, tides and other products of divine manufacture being similarly interpreted.

We ask 'What is its cause?' selectively: we ask it only when we are confronted with some breach of routine, an event that stands out and leads us to ask after its nature and genesis. Thus Sagredo might learn that Galileo was indisposed and ask 'What is the matter?'. On being told that he had cut his arm Sagredo might then ask for the cause of the cut. It is unlikely, however, that on any ordinary Tuesday morning Sagredo would ask after the cause of Galileo's moderate good health. Why should he? Only if he expected Galileo to be otherwise (at a time, for instance, when Padua had been hit by a contagious illness) would the question be in place.

'What caused the clock to stop running?' is a request for news about the one thing responsible for the stoppage, the 'link' immediately preceding cessation of movement.[1] One less frequently asks 'What makes the clock keep running?'. If one does, it is with an awareness that it is different from the former query. No tick-tock account is appropriate here, as it might have been in the former case; a complete account of what makes the clock run will involve a lot of horological theory and physics. (Medical students soon learn the differences between: 1. 'What caused the baby to start breathing?', 2. 'What caused the baby to stop breathing?', and 3. 'What caused the baby to continue to breathe?'.)

Tangibles have become paradigms of what there is; if x can be tripped over, then it exists *par excellence.* This inclines some to reflect on the status of entities whose existence it would be absurd to deny, such as visible objects, facts, causes and effects. The hunt is then on for the tangible, three-dimensional guarantee of their existence. Hence the rendition of seeing and observation as 'the having of a picture'—somewhere. Facts become objects, situations, or states of affairs. Causal sequences become queues of events; and this impedes understanding of physical research. We cannot make the world and science fit a too simple analogy by trimming off the pieces which do not fit.[2]

Natura in reticulum sua genera connexit, non in catenam: homines non possunt nisi catenam sequi, cum non plura simul sermone exponere.

A. VON HALLER (1768)
Historia stirpium indigenarum Helvetiae, vol. II, p. 130.

CHAPTER IV

THEORIES

As soon as we inquire into the reasons *for the phenomena, we enter the domain of theory, which...connects the observed phenomena and traces them back to single 'pure' phenomena, thus bringing about a logical arrangement....* JOOS[1]

A

Typical physical laws are those of motion and gravitation, thermodynamics, electromagnetism, and of the conservation of charge in classical and quantum physics. These were not derived by Bacon's 'Inductio per enumerationem simplicem, ubi non reperitur instantia contradictoria', but some philosophers have thought that they were. A second account treats these laws as high-level hypotheses in a hypothetico-deductive system (hereafter 'H-D' system). It describes physical theory more adequately than did earlier accounts in terms of induction by enumeration, for it says what laws are, and what they can do, in the finished arguments of physicists. But it does not tell us how laws are come by in the first place; and the induction-by-enumeration story at least attempted this.

The two accounts are not alternatives: they are compatible. Acceptance of the second is no reason for rejecting the first. A law might have been arrived at by enumerating particulars; it could then be built into an H-D system as a higher order proposition.[2] If there is anything wrong with the older view the H-D account does not reveal what the fault is.

There *is* something wrong with the older view: it is false. Physicists rarely find laws by enumerating and summarizing observables. There is also something wrong with the H-D account, however. If it were construed as an account of physical practice it would be misleading. Physicists do not start from hypotheses; they start from data. By the time a law has been fixed into an H-D system, really original physical thinking is over. The pedestrian process of deducing observation statements from hypotheses comes only after the physicist sees that the hypothesis will at least explain the initial data requiring explanation. This H-D account is helpful only when discussing the argument of a finished research report, or for understanding how the experimentalist or the engineer develops the theoretical physicist's hypotheses; the analysis leaves undiscussed the reasoning which often points to the first tentative proposals of laws.

The inductive view is that the important inference is from the observation to the law, from the particular to the general. There is something true about this which the H-D account must ignore. Thus Newton wrote: 'The main business of natural philosophy is to argue from phenomena'.[1] The simple inductive view, however, ignores what Newton never ignored: the inference is also from *explicanda* to an *explicans*. The reason for a bevelled mirror's showing a spectrum in the sunlight is not explained by saying that all bevelled mirrors do this. On the inductive account this latter generalization might count as a law: it would accord with Canon Raven's characterization of a natural law as a 'summary of statistical averages'.[2] But only when it is *explained* why bevelled mirrors show spectra in the sunlight will we have a law of the type suggested, in this case Newton's laws of refraction. So the inductive view rightly suggests that laws are got by inference from data. It wrongly suggests that the law is but a summary of these data, instead of being what it must be, an explanation of the data.

H-D accounts all agree that physical laws explain data,[3] but they obscure the initial connexion between data and laws; indeed, they suggest that the fundamental inference is from higher-order hypotheses to observation statements. This may be a way of setting out one's reasons for accepting an hypothesis after it is got, or for making a prediction, but it is not a way of setting out reasons for proposing or for trying an hypothesis in the first place. Yet the initial suggestion of an hypothesis is very often a reasonable affair. It is not so often affected by intuition, insight, hunches, or other imponderables as biographers or scientists suggest. Disciples of the H-D account often dismiss the dawning of an hypothesis as being of psychological interest only, or else claim it to be the province solely of genius and not of logic. They are wrong. If establishing an hypothesis through its predictions has a logic, so has the conceiving of an hypothesis. To form the idea of acceleration or of universal gravitation does require genius: nothing less than a Galileo or a Newton. But that cannot mean that the reflexions leading to these ideas are unreasonable or a-reasonable. Here resides the continuity in physical explanation from the earliest to the present times.

H-D accounts begin with the hypothesis as given, as Mrs Beeton's recipes begin with the hare as given. A preliminary instruction in many cookery books, however, reads 'First catch your hare'.[1] The H-D account tells us what happens after the physicist has caught his hypothesis; but it might be argued that the ingenuity, tenacity, imagination and conceptual boldness which has marked physics since Galileo shows itself more clearly in hypothesis-catching than in the deductive elaboration of caught hypotheses. Galileo struggled for thirty-four years before he was able to advance his constant acceleration hypothesis with confidence. Is this conceptually irrelevant? Will we learn much about Galileo's physical thinking if we just begin our analysis with the constant acceleration hypothesis as a basis for deduction? Was it only the predictions from this hypothesis which commended it to Galileo? The philosopher of science must answer 'No'.

Interpretation is not something a physicist works into a ready-made deductive system: it is operative in the very making of the system. He rarely searches for a deductive system *per se*, one in which his data would appear as consequences if only interpreted physically. He is in search, rather, of an explanation of these data; his goal is a conceptual pattern in terms of which his data will fit intelligibly alongside better-known data. Physics is not applied mathematics. It is a natural science in which mathematics can be applied.

In the thinking which leads to general hypotheses, there are characteristics constant through the history of physics, from Democritus and Heraclitus to Dirac and Heisenberg. Kepler did not *begin* with the hypothesis that Mars' orbit was elliptical and then deduce statements confirmed by Brahe's observations. These latter observations were given, and they set the problem—they were Johannes Kepler's starting point. He struggled back from these, first to one hypothesis, then to another, then to another,[2] and ultimately to the hypothesis of the elliptical orbit. Few detailed accounts have been given by philosophers of science of Kepler's achievements, although his discovery of Mars' orbit is physical thinking at its best. The philosopher of physics should not neglect what Peirce calls the finest retroduction ever made.

When only twenty years of age Kepler became Lecturer of Astronomy at Gratz. The *Prodromus Dissertationum Cosmographicarum continens Mysterium Cosmographicum*[1] appeared two years later, in 1593, and excited Tycho Brahe's interest. They met in Prague in 1600. Brahe was at this time working on his theory of the orbit of Mars, which Kepler found unsatisfactory, since Brahe's miscalculations amounted sometimes to a 5° disagreement with observations (this was not excessive for the geocentric theories then prevalent). Kepler adopted the heliocentric view. Even his earliest reflexions on this surpassed those of Copernicus; for he reasoned that the sun, since it was so near the centre of the planetary system, and so large, must somehow *cause* the planets to move as they do. This was a conjecture of great importance; perhaps the most significant systematic hypothesis yet conceived.

The lines of the apsides of the orbits of Mars and of the earth are not parallel. It seemed to Kepler that the planes of the planetary orbits must intersect, not in the centre of the ecliptic (as all previous astronomers, including Copernicus, had held) but rather in the physical centre of the sun.[2] This assumption affected all of Kepler's work, and consequently all subsequent astronomy, for from it he determined various new methods of calculating the nodes and inclinations of the orbits. These he discovered to be invariable.[3] It appeared also that there was but one reduction from orbit to the ecliptic.

Other features of the idea were disappointing. It led to keen debate with Tycho, and to the unpleasant investigations in the early sections of *De Motibus Stellae Martis*.[4] Kepler thus proposed that the first problem for astronomy should be to master the terrestrial orbit; this before proceeding to other planets. Here was another break with tradition, for in geocentric theories there was of course no terrestrial orbit. Copernicans, moreover, had given the problem only scant attention.

Besides propounding ingenious methods for finding the earth's distance from the sun anywhere in the orbit,[5] Kepler introduced the principle of the bisection. Given (1) that the terrestrial orbit is perfectly circular, (2) that the earth's angular velocity is uniform, and (3) that it requires roughly 360 days to traverse its orbit, then in 180 days it should describe a semi-circle about the sun. Similarly for any larger or smaller interval of time.[1] But these studies disclosed to Kepler that, were the earth's orbit a perfect circle and (at this time there was no question of its being anything else), there could be no fixed point within it about which the planet could describe equal angles in equal times. This principle he had already shown to be of potential explanatory value.

With this improved though still imperfect terrestrial theory, Kepler turned to Mars. The immense collection of observations compiled by Tycho and Longomontanus awaited him. He determined Mars' distances by the same methods as those he had employed in finding the earth's distances from the sun. When these were used to guide his reasoning, however, difficulties were encountered. Kepler's first Martian theory, the 'Vicarious Theory', rested on one principle, fundamental to the astronomical tradition into which he was born. *The planet moved in a perfect circle.* This alone was proper for celestial bodies. They alone exhibited what Aristotle called 'perfect motion'.[2]

However, in these terms the calculated distances required Mars' eccentricity to be very great, so great, in fact, that the resulting equations concerning the orbit's elements were either false or inconsistent. Then Kepler determined these elements by other methods. The result was that the method of equal areas in equal times—on which Kepler was coming to rely—gave errors of 8′ in excess and defect. (Were the orbit really circular this method could not have given errors greater than 1′.)

Kepler was half-inclined to ascribe the errors to imperfections in the method of areas. But he slowly came to suspect that perhaps his predecessors of the previous 2000 years were hasty in thinking the planetary orbits circular. Hindsight makes us underestimate the strength of this ancient maxim; Kepler's challenge seems natural to us. But no bolder exercise of imagination was ever required: Kepler dared to 'pull the pattern' away from all the astronomical thinking there had ever been. Not even the conceptual upsets of our century of natural science required such a break with the past. Before Kepler, circular motion was to the concept of planet as 'tangibility' is to our concept of 'physical object'. If intangible physical objects are inconceivable to us, so were non-circular planetary orbits to Kepler's predecessors, *and* contemporaries. Remember, Tycho and Galileo never made this break.

Kepler gave three tentative arguments for the view that the Martian orbit might not be circular.[1]

1. From the supposition that the orbit was circular Kepler calculated the longitude of the aphelion, the eccentricity of the orbit, and its ratio to the terrestrial orbit. These were irreconcilable with the observed elements. Moreover, the results of any one combination of distances derived by these calculations were inconsistent with the results of other combinations of distances. 'Nor do the equations computed physically agree with the observed facts (the equations are represented here by the vicarious hypothesis).'[2]

2. More significantly, a calculation of three carefully observed distances (at 10°, 104° and 37° of arc from aphelion), when plotted against a circular orbit revealed that the plane actually retired within the circular path by 350, 783 and 789 parts in 100,000. 'What is to be said?...The planet's orbit is not circular, but it recedes at either side slightly, and returns to the amplitude of the circle at perigee; a figure describing such a path is called an oval.'[3] Mar's orbit, then, may not be circular. It may consist in a curve which coincides with the circle at the apsides, and then retires within it; at 90° and 270° of eccentric anomaly it will deviate most from the circle.

3. In attempting to derive the equations for an eccentric circular orbit by the method of areas, the area of the circle was assumed to equal the whole time of the planet's revolution. Any sector of this circle was thus taken to be equivalent to the time taken by the planet to describe that sector's particular arc. Were the orbit not in fact a circle, however, errors committed through using circular sectors to measure the times taken traversing the arches would be unavoidable. At 90° and 270°, the method of areas would give the times as too long and the planet's motion as too slow. No such errors would arise with a curve for Mars which retires within the circle. 'From which it is shown what I promised to do in chapters XX, XXIII...that the planet's orbit is not a circle but has the figure of an oval.'[4]

Peirce records[5] that Kepler now proved that the circle was compressed. On the contrary, Kepler's first reaction to his own tentative arguments was again to question his method of areas. He still argued thus: given that the orbit is circular, and supposing my reasoning to be correct, then the observations α, β, γ, are directly predicted; but α, β and γ do not occur; *therefore my reasoning was not correct.* After failing to reconcile the circular orbit with the equations given by the method of areas, he actually abandoned *the latter.* Different considerations were required to convince him that it was the circular orbit hypothesis which was spoiling his theory. Only when the distances given to him by the circle were repeatedly inconsistent with those observed by Tycho, did Kepler begin systematically to doubt the circular orbit hypothesis. Even then he headed the next chapter[1] 'De Causis Naturalibus Hujus Deflexionis Planetae A Circulo'. However, after his inquiry there he concluded:

> Consider, thoughtful reader, and you will be transfixed by the force of the argument. For I could not think of any other means of making the orbit of the planet ovoid. As these things presented themselves to me in this way, the magnitude of this recession at the sides being securely established, as well as the agreement of the numbers, I celebrated another Martial triumph.

And even

> And we, good reader, will not indulge in this splendid triumph for more than one small day... restrained as we are by the rumours of a new rebellion, lest the fabric of our achievement perish in excessive rejoicing.[2]

Kepler fits the non-circular curve which coincides with the circle at the apsides; these two points are accurately determined. The new curve retires within the circle at 90° and 270° by 858 parts of its semi-diameter (supposed to contain 100,000). This is the famous *figuram ovalem,* whose role has been misunderstood by many philosophers and historians who have considered Kepler's own account.[3] We must proceed with care and in detail: these are crucial moments in the long and ultimately triumphant retroduction.

One thing cannot be overstressed, Kepler's first non-circular curve was not itself an ellipse.

> Whichever of these ways is used to describe the line on which the planet moves, it follows that this path, indicated by the following points, δ, μ, γ, σ, π, ρ, λ, *is ovular, and not elliptical*; to the latter, Mechanicians wrongly give the name derived from *ovo.* The egg (ovum) can be spun on two vertices, one flatter (obtuse), one sharper (acute). Further it is bound by inclined sides. This is the figure I have created.[4]

Compare:

All of this conspires to show that the resegmentum of our eccentric circle is much larger below than above, in equal recession from the apsides. Anyone can establish this either by numerical calculation or by mechanical drawing—some eccentricity being assumed.[1] (See fig. 9.)

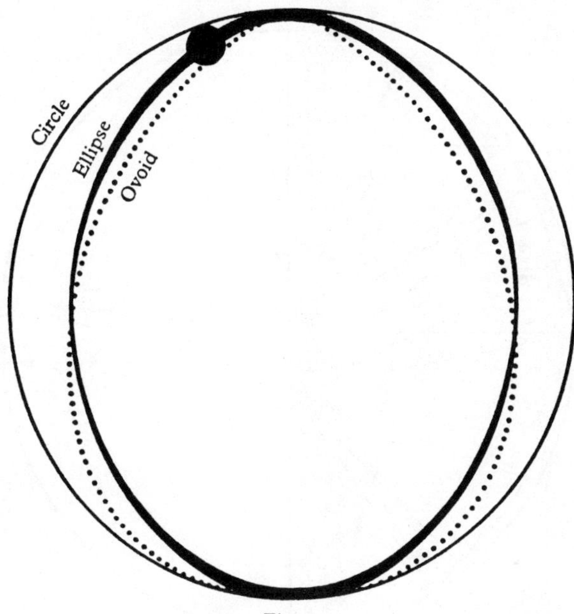

Fig. 9

Commentators are almost unanimous in thinking that Kepler's first departure from the circle was to an oval: that is, a perfect ellipse;[2] according to them this first curve was an ellipse, only one of the wrong dimensions. But in fact it was not an ellipse at all: it was, rather, a 'plani oviformis'.[3] There is some reason for this misinterpretation; Kepler's own confusion is partly responsible. It has always seemed odd to me that Kepler should have jumped from the circle straight to the usually-reported ellipse without seeing immediately the solution to all his problems. (We exercised care when distinguishing the mathematical and physical aspects of Beeckman's conservation principle. Here too we must distinguish Kepler's physical hypothesis, namely that Mars describes an oviform around the sun, from his mathematical hypothesis, which involved calculations with a perfect ellipse.)

The chronology of Kepler's conceptual development is more intelligible in this perspective. For in all Martian theories up to and including Kepler's there was but one focus for the orbit. The

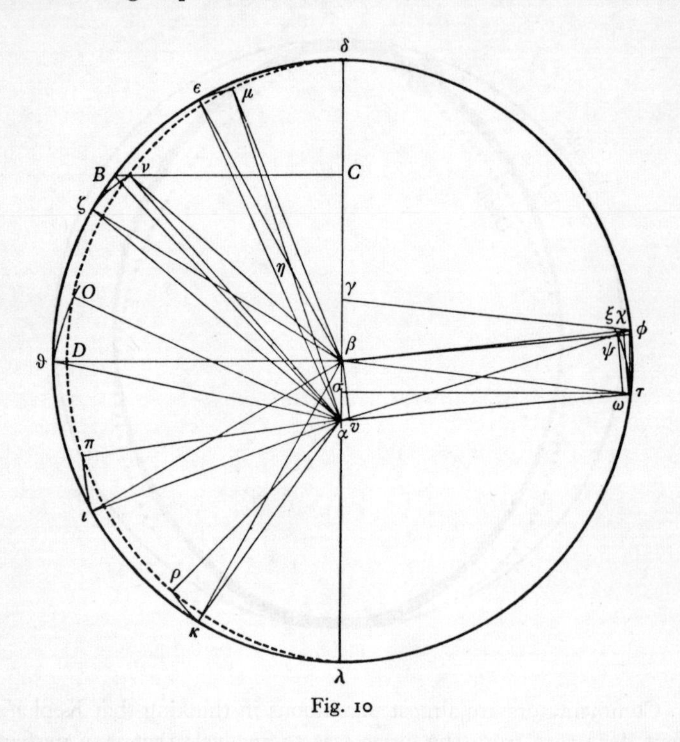

Fig. 10

circle, even with the sun or earth placed eccentrically within it, gave a system with one geometrical focus. Naturally, if one is going to depart from the dictum of a circular orbit, the conceptual strain in doing so will be less if the curve substituted itself has only one focus. The oviform cannot have two foci. That is ultimately what gave Kepler all his trouble, for he could work out the properties of this queer geometrical entity only by considering it as an approximation to a perfect ellipse, whose properties were explored by Archimedes. Only indirectly could Kepler learn anything useful about this recalcitrant figure. *The move of treating observed physical phenomena as approximations to mathematically 'clean' conceptions developed after Kepler into a defining property of physical inquiry.*

What a fortunate accident! Had the ovoid been tractable, Kepler might never have had to introduce the ellipse into his thinking at all, not even as a geometrical prop. He might never have deserted the one-focus curve. Kepler's exposition indicates that his having had to introduce this prop made the later physical hypothesis of an elliptical Martian orbit (with the sun in one of the two foci) more plausible than it ever could have been had he been dealing exclusively with one-focus curves.[1]

At first Kepler made fruitless attempts to find directly the quadrature of the oviform curve; without this no equations could be forthcoming. But he then conjectured that were the ovoid supposed to be sensibly equal to an ellipse of the same eccentricity (and described by the same greater axis), the lunula cut off by it would be but insensibly different from that cut off by a perfect ellipse. Thus he proceeded: '*If* our figure were a perfect ellipse...', and 'Let us suppose then (or let it be given that) our figure is a perfect ellipse, from which it differs little. Let us see what follows therefrom.'[2] Compare: 'The general geometrical properties of the perfect ellipse are manifested in the actual ovoid curve, from which it is but insensibly different, since the defects above almost exactly compensate the excesses below....'[3]

Nonetheless it was still impossible to find the oviform's equations by the method of areas, which had now regained Kepler's confidence. Kepler calls on the geometers 'eorumque opem imploro'. Every time the physical ovoid is treated as an approximation to the geometrical ellipse, Kepler's calculations put the sun in one focus of the latter. Thus '*in distantia mediocri Planetae* τ a Sole α'[4] and '*a circumferentia* ζ *versus Solem* α'.[5] This is profoundly important. It was years later in Kepler's research, and 100 pages further on in *De Motibus Stellae Martis*, that Kepler came to treat the ellipse with the sun in one focus as a physical hypothesis describing Mars' orbit.

The two diagrams, figs. 10 and 11, are essentially similar. Their impacts on Kepler, however, were as totally different as the bird is from the antelope, the hidden man from the cluster of patches in which he hides, Galileo's dawn from Simplicius', Beeckman's problem from Descartes', Leverrier's conception of Mercury from that of Einstein, and Schrödinger's 'ψ' from Born's. The earlier ellipse (fig. 10) is a formal prop for Kepler's physical thinking, just as the binomial expansion often serves today in quantum field theory. The later ellipse (fig. 11) holds the real key to the Keplerian systematization of Brahe's data. But the earlier diagram made the later one possible.

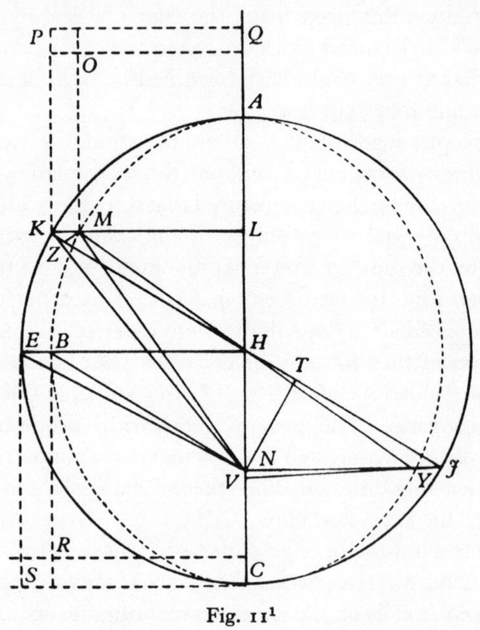

Fig. 11[1]

Such are the delicate conceptual threads that get tangled or broken if one identifies the physical oviform with a geometrical ellipse, or if one separates them too sharply. Midway in the march of Kepler's thoughts, the ellipse is 'there'. But it is there differently from the manner of its presence later; its organization is different.

This early oviform theory gave the motions of Mars as too slow about the apsides, and too quick about the mean distances. These were errors contrary to those detected in a circular orbit. It was uncertain whether these arose from the theory or from imperfections in the expression of its principles. But it did appear that a mean of the circular and the elliptical areas might approach the true orbit. Kepler noted this, but nonetheless tried to deduce the equations of Mars directly from the distances given by the oviform curve: a loathsome undertaking. (The relevant passages in *De Motibus Stellae Martis* are all perspiration with little inspiration.) But Mars was still described too slowly about the apsides, and too rapidly about the mean distances. Increasing the eccentricity, a move that was often effective in other cases only increased these errors. Moreover, the ovoid curve which Mars was supposed to trace had to be divided into unequal portions, because of its geometry. The curvature and length of these arcs were discovered to be inversely as their distance from the sun. By the theory, however, only the solar force acts in inverse ratio of the distances; the planet's intrinsic force is invariable. This is inconsistent with the ovoid's differing curvatures and lengths of arc. (As Galileo argued against the impetus theorists, one cannot get a variable effect from a constant cause.) The inconsistency led Kepler to doubt the accuracy of his measurements of the oviform orbit, construed as the joint product of the Martian and the solar force.

Kepler was induced to abandon further attempts to obtain the ovoid's quadrature. As a result of a comparison between the distances given by the oviform and the corrected observations taken by Tycho (at forty different points of anomaly), it became clear that the real orbit lay between the circle and the ovoid, his approximate ellipse. In his own words:

> My reasoning was the same as that of Chapters XLIX, L, and LVI. The circle of Chapter XLIII errs by excess. The ellipse of Chapter XLV errs by deficiency. And the excess and the deficiency are equal. There is no other middle term between a circle and an ellipse, save another ellipse. Therefore the path of the planet is an ellipse....[1]

The original text has '*ellipsis capitus XLV*...'; this suggests why later commentators have almost always gone wrong. For the curve discussed in ch. XLV is oviform; it allows the physical hypothesis of the oviform orbit, which Kepler dealt with as an approximation to a formal ellipse. Now he refers to that hypothesis as having *been* an ellipse.[2] Like most physicists caught up in genuine problems, he was not clear himself about the physical and mathematical aspects of these 'frontier' hypotheses.

To return: the only curve which would explain the data had to be 'another' ellipse. Kepler now stumbled upon an important correlation.[1] The 660 parts of a semi-diameter equalling 152,350 were equivalent to 432 parts of a semi-diameter equalling 100,000. This is nearly the number 429, which is exactly one-half of the 858 he had found to be the extreme breadth of the lunula cut off in the oval theory.[2] Attending now to the greatest optical equation of Mars (which is between 5° 18′ and 5° 19′) he saw that 429 was also the excess of the secant of 5° 18′ above the radius 100,000.[3] By an impressive argument he showed that the distances used in the circle were the secants of the optical equations in all points of eccentric anomaly. If, instead of these, he used the different *radii* to which they were the secants, the resulting calculated distances would agree with those observed.

Alas, here Kepler blundered in his calculation of the planet's positions at determined distances. Truth seemed very reluctant to deliver herself up to Kepler and he applied to her Virgil's

> Malo me Galataea petit, lasciva puella,
> Et fugit ad salices, et se cupit ante videri.[4]

Again he failed; again Tycho's data resisted the pattern he proposed. His distances, though observationally accurate, were inconsistent with the elliptical form he ascribed to the orbit. The observed distances would have required a new oviform figure, again exceeding the ellipse in the first and fourth quadrants and retiring within it in the second and third. Kepler ascribed the error to his ancillary theory of librations, wherein Mars was supposed to oscillate at right angles to its orbit through its whole revolution. Though this gave accurate distances, he abandoned it.

Now, with little conviction, Kepler returned to the ellipse. This seemed the only way of preserving his other conceptually valuable principles. He supposed that in doing this he was appealing to something totally different from his theory of librations.[5] But slim hopes of success were held: 'It is clear therefore that the path has cheeks; it is not an ellipse. And while an ellipse offers just equations, this cheeky figure offers unjust ones.'[6] The reconsideration of the ellipse was thus something into which Kepler retreated after finding no other prospect for applying the principles he had established. The ellipse as a physical hypothesis began to beckon. But now Kepler became worried to distraction through trying to understand why Mars should abandon (in favour of an elliptical path) the librations, on whose assumption accurate calculations of distance were produced. He toiled on this problem like a man possessed, until finally his perplexities dissolved before an insight which transformed his data and all subsequent astronomy and physics. In his own words:

Yet even this was not the crux of the matter. Indeed *the* great problem was this: that in considering and casting about to the very limits of my sanity, I could not discover why the planet, to which with such great probability and agreement of observed distances the libration *LE* in diameter *LK* could be attributed, why the planet should prefer an elliptical path indicated by the equations.

Oh how ridiculous of me! As though the libration in diameter could not lead to the elliptical path. This idea carried no little persuasive force —the ellipse and the libration stand or fall together, as will be made clear in the next Chapter. There it will be demonstrated that there is no figure of a planet's orbit other than the perfect ellipse—the concurrence of reasons drawn from the principles of physics, with the experiences of observations and hypotheses being adduced in this chapter by anticipation.[1]

This is a model of differences in conceptual organization. Before 'O me ridiculum!' the libration theory and the elliptical theory were distinctly different for Kepler. The difference between 'librations *v.* ellipse' and 'librations = ellipse', is like the difference between the bird and the antelope, or the bear on the tree before and after identification.

The arithmetic blunder was quickly discovered, and the discovery led to the fullest confirmation of the hypothesis.[2] As visual phenomena of ch. I were cast into patterns by the appreciation of some particular dot or line, so here the enormous heap of calculations, velocities, positions and distances which had set Kepler his problem now pulled together into a geometrically intelligible pattern. The elliptical areas were seen to be equivalent; similarly, the sums of the corresponding diametral distances were equal; the equations following from the ellipse were general expressions of Tycho's original data. All this made it clear that Mars revolves around the sun in an ellipse, describing around the sun areas proportional to its times.

Predictions to unobserved positions of Mars had not yet been undertaken. Nor did Kepler feel it absolutely necessary to endure this before proposing the elliptical orbit hypothesis.

This was a physical discovery. Since the same physical conditions obtained throughout the solar system, the same equations ought to explain other planetary revolutions as well. These three great *explicantia* are the well-known result: (*a*) that planetary orbits are elliptical with the sun in their common focus (1609), (*b*) that they describe around the sun areas proportional to their times of passage (1609), and (*c*) that the squares of the times of their revolutions are proportional to the cubes of their greater axes, or their mean distances from the sun (1619). These are most important in the history of astronomy. They supplied the material for Newton's retroduction to the law of universal gravitation.[1]

Of this monumental reasoning from *explicanda* to *explicans* could any account be more ludicrous than that of J. S. Mill, who argued that Kepler's law is just 'a compendious expression for the one set of directly observed facts'?[2] Mill had no real experience of theoretical astronomy. But he might have perused *De Motibus Stellae Martis*. Whewell is rightly uneasy about Mill's account.[3] His alternative account, however, turns on Kepler's having got the hypothesis as a 'colligating concept'. This is little better than the modern hypothetico-deductive account which has it that Kepler succeeded 'by thinking of general hypotheses from which particular consequences are deduced which can be tested by observation'.[4]

Kepler typifies all reasoning in physical science. Would it have required so much time, and genius, to 'observe' the elliptical orbit in Tycho's data? *De Motibus Stellae Martis* is more than a compendious expression of Brahe's observations. Nor is it concerned with deducing geometrical consequences from the elliptical orbit hypothesis, 'thought of' Kekulé-fashion. Kepler's task was: given Tycho's data, what is the simplest curve which includes them all? When he at last found the ellipse his work as a creative thinker was virtually finished. Any mathematician could then deduce further consequences not included in Tycho's lists. It required no genius to take Kepler's idea and try it for other planets.

Kepler never modified a projected explanation capriciously; he always had a sound reason for every modification he made. When he did make an adjustment which exactly satisfied the observations, it stood 'upon a totally different logical footing from what it would if it had been struck out at random...and had been found to satisfy the observations. Kepler shows his keen logical sense in detailing the whole process by which he finally arrived at the true orbit. This is the greatest piece of Retroductive reasoning ever performed.'[1]

B

What features of inference by retroduction does Kepler bring out? The reasoning from surprising data to an explanation binds together as 'physics' centuries of inquiry, methods, techniques and problems.

Was Kepler's struggle up from Tycho's data to the proposal of the elliptical orbit hypothesis really inferential at all? He wrote *De Motibus Stellae Martis* in order to set out his reasons for suggesting the ellipse. These were not deductive reasons; he was working from *explicanda* to *explicans*. But neither were they inductive—not, at least, in any form advocated by the empiricists, statisticians and probability theorists who have written on induction.[2]

Aristotle lists the types of inferences. These are deductive, inductive and one other called 'ἀπαγωγή'. This is translated as 'reduction'.[3] Peirce translates it as 'abduction' or 'retroduction'. What distinguishes this kind of argument for Aristotle is that

> the relation of the middle to the last term is uncertain, though equally or more probable than the conclusion; or again an argument in which the terms intermediate between the last term and the middle are few. For in any of these cases it turns out that we approach more nearly to knowledge...since we have taken a new term.[4]

After describing deduction in a familiar way, Peirce speaks of induction as the experimental testing of a finished theory.[5] Induction

> sets out with a theory and it measures the degree of concordance of that theory with fact. It never can originate any idea whatever. No more can deduction. All the ideas of science come to it by the way of Abduction. Abduction consists in studying facts and devising a theory to explain them. Its only justification is that if we are ever to understand things at all, it must be in that way. Abductive and inductive reasoning are utterly irreducible, either to the other or to Deduction, or Deduction to either of them....[6]

> Deduction proves that something *must* be; Induction shows that something *actually is* operative; Abduction merely suggests that something *may be*.[7]

> ...man has a certain Insight, not strong enough to be oftener right than wrong, but strong enough not to be overwhelmingly more often wrong than right....An Insight, I call it, because it is to be referred to the same general class of operations to which Perceptive Judgments belong....If you ask an investigator why he does not try this or that wild theory, he will say 'It does not seem *reasonable*'.[1]

Peirce regards an abductive inference (such as 'The observed positions of Mars fall between a circle and an oval, so the orbit must be an ellipse') and a perceptual judgment (such as 'It is laevo-rotatory') as being opposite sides of the same epistemological coin. *Seeing that* is relevant here. The dawning of an aspect and the dawning of an explanation both suggest what to look for next. In both, the elements of inquiry coagulate into an intelligible pattern. The affinities between seeing the hidden man in a cluster of dots and seeing the Martian ellipse in a cluster of data are profound. 'What can our first acquaintance with an inference, when it is not yet adopted, be but a perception of the world of ideas?'[2] But '...abduction, although it is very little hampered by logical rules, nevertheless is logical inference, asserting its conclusion only problematically, or conjecturally, it is true, but nevertheless having a perfectly definite logical form'.[3]

Before Peirce treated retroduction as an inference[4] logicians had recognized that the reasonable proposal of an explanatory hypothesis was subject to certain conditions. The hypothesis cannot be admitted, even as a tentative conjecture, unless it would account for the phenomena posing the difficulty—or at least some of them. This is understressed in most H-D accounts of physical theory, and it is non-existent in simple inductive accounts. The form of the inference is this:

1. Some surprising phenomenon P is observed.
2. P would be explicable as a matter of course if H were true.
3. Hence there is reason to think that H is true.

H cannot be retroductively inferred until its content is present in 2. Inductive accounts expect H to emerge from repetitions of P. H-D accounts make P emerge from some unaccounted-for creation of H as a 'higher-level hypothesis'.

'Mars' positions would fall between a circle and the oviform as a matter of course if its orbit were elliptical'; 'the distance dropped by a body would be $\frac{1}{2}at^2$ as a matter of course if the acceleration of a freely falling body were constant'. The *H*'s here did not result from any actuarial or statistical processing of increasingly large numbers of the *P*'s. Nor were they just 'thought of', the *P*'s being deducible from them.[1]

Perceiving the pattern in phenomena is central to their being 'explicable as a matter of course'. Thus the significance of any blob or line in earlier diagrams eludes one until the organization of the whole is grasped; then this spot, or that patch, becomes understood as a matter of course. Why does Mars appear to accelerate at 90° and 270°?—(*P*). Because its orbit is elliptical—(*H*). Grasping this plot makes the details explicable, just as the impact of a weight striking clay becomes intelligible against the laws of falling bodies. This is what philosophers and natural philosophers were groping for when they spoke of discerning the nature of a phenomenon, its essence;[2] this will always be the trigger of physical inquiry. The struggle for intelligibility (pattern, organization) in natural philosophy has never been portrayed in inductive or H-D accounts.

Consider the bird-antelope in fig. 12. Now it has additional lines. Were this flashed on to a screen I might say 'It has four feathers'. I may be wrong: that the number of wiggly lines on the figure is other than four is a conceptual possibility. 'It has four feathers' is thus falsifiable, empirical. It is an observation statement. To determine its truth we need only put the figure on the screen again and count the lines.

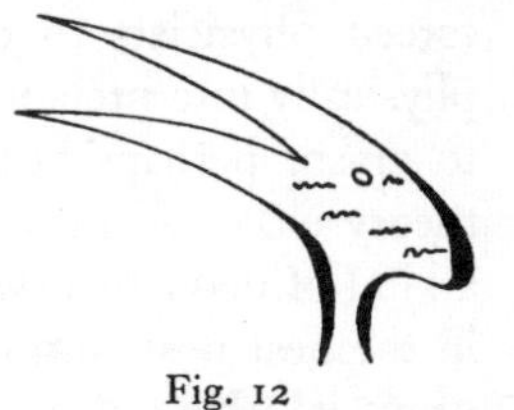

Fig. 12

The statement that the figure is of a bird, however, is not falsifiable in the same sense. Its negation does not represent the same conceptual possibility, for it concerns not an observational detail but the very pattern which makes those details intelligible. One could not even say 'It has four feathers' and be wrong about it, if it was not a feathered object. I can show you your error if you say 'four feathers'. But I cannot thus disclose your 'error' in saying of the bird-antelope that it is a bird (instead of an antelope).

Pattern statements are different from detail statements. They are not inductive summaries of detail statements. Still the statement, 'It's a bird' is truly empirical. Had birds been different, or had the bird-antelope been drawn differently, 'It's a bird' might not have been true. In some sense it is true. If the detail statements are empirical, the pattern statements which give them sense are also empirical—though not in the same way. To deny a detail statement is to do something within the pattern. To deny a pattern statement is to attack the conceptual framework itself, and this denial cannot function in the same way.

P and *H* must have further logical properties in order to figure in '*P* would be explicable as a matter of course if *H* were true'. If *H* is meant to explain *P*, then *H* cannot itself rest upon the features in *P* which required explanation. This is why the peculiar colour and odour of chlorine (*P*) are not explained by reference to atoms in a volume of chlorine, each one having the colour and odour in question (*H*). Grasping this point is essential for any understanding of the fundamental concepts of modern particle physics.[1] This feature of retroductive reasoning shows why elementary particles must be unpicturable; why all electrons must be identical; why the 'state' of a proton cannot be determined precisely; why recent attempts to rectify particle theory have necessarily forced physicists to consider matter as lacking in any direct, physically interpretable properties. These things philosophers fail to grasp, perhaps because they are inclined to regard physical theory either as an inductive compound on the one hand, or as a kind of deductive system on the other. Of all men Kepler was in the best position to say 'Mars has no unique orbit, and I can prove it'. He did not say this. Galileo could have said $s=\frac{1}{2}at^2$, and no more, but he pressed on. Newton pressed on; Einstein, Bohr, De Broglie, Schrödinger, Heisenberg and Dirac pressed on —for explanations, which no amount of statistical repetition or deductive ingenuity alone could ever supply.

The critical moment comes when the physicist perceives that one might reason about the data in such and such a way. One might explain this welter of phenomena *P*, throw it all into an intelligible pattern, by supposing *H* to obtain. But *P* controls *H*, not vice versa. The reasoning is from data to hypotheses and theories, not the reverse.

> Retroduction...begins always with colligation, of course, of a variety of separately observed facts....How remarkable it is...that the entire army of logicians...should have left it to this mineralogist [Whewell] to point out colligation as a generally essential step in reasoning.

Abduction...amounts...to observing a fact and then professing to say what...it was that gave rise to that fact....[1]

Kepler's was a great retroduction. Galileo's discovery that gravitational acceleration is constant was another. We left him in 1604 with the wrong hypothesis about freely falling bodies. He had told Sarpi that the velocities of a falling body were proportional to the distances it had fallen. By 1609 Galileo had realized his mistake, and was arguing that the body's velocities were proportional not to the distances fallen but rather to the times of fall. This was an innovation of great importance; for the relation between a velocity parameter, $v \propto t$, is not a 'natural' one in the way that '2 gm. plus 2 gm. equals 4 gm.' is. This is already a step towards the modern situation, wherein a theoretical physicist must be expert in the theory of functions. Triumphs in contemporary physics consist in discovering that one parameter can be regarded as a function of some other one. The 'real' physical relation between them may be unobvious or non-existent. In his hypothesis of 1609 Galileo's feet are on this path: he is pursuing a prior *explicans*, one having something to do with acceleration. Galileo's thirty-four-year march towards his final explanation is punctuated with misconceptions and erroneous arguments which it would be instructive to re-examine, but the matter cannot be pursued here.[2] Suffice it to say that he always tries to explain his original data by fashioning general hypotheses and theories 'in their image'. His hypotheses are never inductive summaries of his data; nor does he actively doubt them until he can deduce new observation statements which experiments confirm. Galileo knew he had succeeded when the constant acceleration hypothesis patterned the diverse phenomena he had encountered for thirty years. His reasoned advance from insight to insight culminated in an ultimate physical *explicans*. Further deductions were merely confirmatory; he could have left them to any of his students—Viviani or Toricelli. Even had verification of these further predictions eluded seventeenth-century science, this would not have prevented Galileo from embracing the constant acceleration hypothesis, any more than Copernicus and Kepler were prevented from embracing heliocentrism by the lack of a telescope with which to observe Venus' phases. Kepler needed no new observations to realize that the ellipse covered all observed positions. Newton required no new predictions from his gravitation hypothesis to be confident that this really did explain Kepler's three laws and a variety of other given data.

Physical theories provide patterns within which data appear intelligible. They constitute a 'conceptual Gestalt'.[1] A theory is not pieced together from observed phenomena; it is rather what makes it possible to observe phenomena as being of a certain sort, and as related to other phenomena.[2] Theories put phenomena into systems. They are built up 'in reverse'—retroductively. A theory is a cluster of conclusions in search of a premiss. From the observed properties of phenomena the physicist reasons his way towards a keystone idea from which the properties are explicable as a matter of course. The physicist seeks not a set of possible objects, but a set of possible explanations.

Some general remarks about causal theories will lead us to a discussion of Classical Particle Theory. Ch. III stressed the theory-backed character of causal talk. The necessity sometimes associated with event-pairs construed as cause and effect is really that obtaining between premisses and conclusions in theories which guarantee inferences from the one event to the other. This is masked by our causal idioms. Causal inferences are rarely set out in explicit demonstrations; they are built into the meanings of words in certain contexts. 'Steering gone!', 'Puncture!', 'Insulation leak!', 'Saturated!', can be uttered in ways so pregnant causally that inferences to subsequent events seem inevitable.

Contiguity, propinquity, regular succession: these are properties of certain kinds of events which could form the subject matter of a physical theory. A causal theory is just a theory which guarantees inferences between events of this kind. But naturally not all events are of this kind; not all physical theories are causal theories.

Event-pairs which are contiguous, propinquitous and regularly successive, but not causally related, are well known: they are called coincidences and 'merely' statistical regularities. No theory directly binds these events. What philosophers sought as the objective necessity of causal sequences resides in the form of the theory which connects descriptions of these sequences.

Even so, there is probably no fixed idea of what a causal theory is. The concept varies with the impact particular theories have on us. When the *Principia* was published, Newton's theory of motion was regarded as abstract, merely mathematical. Huygens and Leibniz never thought the theory a satisfactory physical explanation—that is, a causal account—of particle motion and interaction.[1] The *Principia* was but a formula. Galileo's recantation, remember, consisted in conceding that his doctrine of the earth's motion was correct only as mathematical fiction, but false as a causal account of planetary and solar behaviour.

Newton himself felt the force of this distinction. With his contemporaries, and with his Cartesian adversaries, he judged that the abstract formalism of the *Principia* needed mechanical supplementation to satisfy the need for a casual explanation of phenomena. With the dictum 'Hypotheses non fingo', however, he declined to satisfy this need.[2]

As the *Principia* won over the physicists' thinking, however, it became a model for every other field of inquiry. Soon it ceased being a merely mathematical aid to the prediction of how bodies behave, and became a system, indeed *the* system of mechanics. The word 'mechanistic' was used to mark processes which permitted explanation in terms of the *Principia.* Newton gave new meaning to causal explanation. Mechanics became the paradigm of a causal theory. Thus its central concepts—force, mass, momentum —were sometimes regarded as ultimate causal powers. They became rather like impetus in pre-Galilean thought; if an explanation could be traced back through this system to one or several of the laws of motion (in which these concepts figure), that explanation was causal in nature.

This attitude was expressed by Helmholtz: 'To understand a phenomenon means nothing else than to reduce it to the Newtonian laws. Then the necessity for explanation has been satisfied in a palpable way.'[3]

Professor Broad writes (in 1913): 'The laws of mechanics give rise to much the most certain and important physical science that exists, and so they are good examples of causal laws at their best.'[4]

'Causal' and 'mechanical' became identified with 'picturable'. It was hinted that the criterion for determining whether a physical theory was causal or mechanical was whether it could be pictured. *Principia* became the archetype of picturability. Maxwell's field equations for electrodynamical phenomena were criticized for not being picturable. Many attempts were made to supplement his theory with a mechanical model.[5]

In elementary particle theory today phenomena are 'encountered' which are neither causal, nor picturable, nor even mechanical in any classical sense. For example, the theory requires that the nucleus of every unstable isotope be identical with every other nucleus of that type—in a stronger sense of 'identical' than anything yet encountered in physics. But these nuclei decay in an unpredictable way (another part of the theory requires that); so the decay cannot be conceived of as a caused event. For the nuclei of those atoms of carbon 14 which *do* decay are internally identical, until the instant of decay, with those which do not decay. This leads physicists to say unrepentant things about the collapse of the law of causality in modern science. Yet elementary particle theory makes a fountain of diverse observations intelligible; indeed, our ideas of picturability and causality are already broadening.[1] Our ideas of the nature of a mechanical system are broadening too. But before turning to this, let us see how classical mechanics patterned the Victorian physicists' thinking about nature.

CHAPTER V

CLASSICAL PARTICLE PHYSICS

...the fact that it can be described by Newtonian Mechanics tells us nothing about the world; but this *tells us something, namely, that the world can be described in that particular way in which as a matter of fact it is described.* WITTGENSTEIN[1]

Laws of classical particle physics exercise philosophers. Statements of these laws are in some sense empirical, yet they seem often to resist the idea of disconfirmation: evidence against them is sometimes impossible to conceive.

Newton stressed the empirical basis of dynamics: Broad has inherited this interest in the evidence supporting dynamical law statements; he regards them as substantive, descriptive, empirical propositions. Myriad confirmations[2] and a central place in the system of dynamical concepts—these are the only reasons why it is difficult to imagine a macrophysical world in which the laws do not obtain.

But other thinkers are impressed by the resistance of dynamical law statements to falsification. Poincaré is typical of those who regard such statements as conventions, or definitions, or procedural rules, or boundary conditions.[3] Hence, for him it is their empirical aspects that must be explained, or explained away.

Seen through classical dichotomies, classical mechanics is challenging. It springs from empirical propositions against which disconfirmation is not always conceivable. Disconfirmation would result not in conceptions which negate those in the law statements, but in no coherent conceptions at all. Apparently we must explain away either their conventional aspects or their contingent features; they are not to float betwixt and between. (Kant refused to explain away either. He was in some ways a better observer of physics than his critics; for him, being betwixt and between was the virtue of Newton's dynamics.)

So much for the celebrated question 'What is the logical status of the laws of classical particle physics?', to which Broad, Poincaré and Kant have given important, but single-valued answers. The question itself is misleading. It is like asking 'What is *the* use of rope?'. The replies to this are no fewer than the uses for rope. There are as many uses for the sentences which express dynamical law statements as there are types of context in which they can be employed. In trying to provide *the* answer to the above query, Broad, Poincaré and Kant (not to mention Mill, Whewell, Mach, Pearson, Russell, Braithwaite and Toulmin) have shown how versatile physicists really are with the sentences and formulae of dynamics. There is no such thing as *the* law of inertia, *the* law of force, *the* law of gravitation.

Let us contrast the actual scientific uses of dynamical law statements with philosophical commentaries on their status. This will raise questions about the relationship between the uses of law sentences and the logic of law statements, about the ways in which the latter can be regarded as *a priori*, and about other matters.

A

Dynamical law statements help to explain physical events. An event is explained when it is traced to other events which require less explanation; when it is shown to be part of an intelligible pattern of events.

On striding into my study I slide abruptly across the floor—it has just been polished. There is no more to say, no need for further explanation. This means, not that there is no further explanation, but only that it is too obvious; the effect on perambulation of polished floors is no secret. That my floor has been polished is all the explanation needed for this performance: the general reason why shoe leather slips on polish is not of immediate relevance in explaining why I slipped. Trace an event to incidents which are commonplace and we are rarely interested in tracing it further. The pattern is too clear.

When events have been explained by linking them to statements of the laws of classical particle physics, however, this cannot be because the *explicans* is commonplace. Aristotle was able to detect the commonplace, yet he would have denied at least part of the first law of motion, namely: 'All bodies remain either at rest or in uniform rectilinear motion, unless compelled by impressed forces to change their state.'[1] This is not obviously true; the Philosopher treats it as clearly false.[2]

But some Newtonians felt that this law statement explained events not because it expressed a commonplace, but because *it* needed no explanation. That the area within a circle is the maximum for any closed curve of that perimeter needs no explanation; that is what circles are. The quotation above sets out how bodies do move; what further is there to explain? Kepler's discovery left nothing further to explain about Mars' motion. Dynamical explanations derive from statements like this one above. Why expect that this statement can itself be explained dynamically?[1]

So a statement of the law of inertia describes a kind of event (inertial motion) whose explanation, while not obvious, is not as a matter of principle required. What is to be said of this comment on the law of inertia? Some events need less explaining than others. That the Earth moves needs less explaining than that it moves in an ellipsoidal orbit and rotates on its axis; these latter require all the explaining needed by the former, and more besides.

If that to which I refer when accounting for events needs more explaining than that to which you refer, then your explanation is better than mine. Kepler's astronomy needed less supplementary explanation than Tycho's, and so was better. Galileo's cosmology required less explanation than did that of his Ptolemaic adversary, Simplicius: therefore Galileo's was better.[2] Because Aristotle's account of the natural motion of bodies required more *ad hoc* explanation than the account in the first law of motion, Newton's was better.

Apparently, then, the best explanation must show how an event needing no explanation (inertial motion) is connected with observed events. But this makes it seem that the goal of physics is to explain the contingent in terms of the *a priori*; to account for events needing explanation in terms of those which need none at all. The goal seems to be to relate vulnerable statements with those which are invulnerable. This view is not absent from the history of mechanics. Latter-day Newtonians regarded dynamical law statements as needing no explanation whatever;[3] for various reasons they were treated as prescriptive, immutable, *a priori*. Indeed, this is what such statements seemed designed to be—the ultimate shackles in chains of physical explanation.[4] Many physicists have used them thus.

Apparently a statement of the first law needs no explanation, because it could not be false; yet it tells us what happens in nature, or what would happen if certain conditions were realized. Thus the empirical grounds for asserting the first law are events like slipping on polished floors, or observing how a round rock moves across ice with but slightly diminishing velocity until it slows to a halt. When the first law statement seems not to hold, the reason can always be found: ground glass on the ice, perhaps, or the discovery that the rock is a lodestone, etc. The law encapsulates and extrapolates much information about events, yet it seems beyond disconfirmation: it could not but be true.

'But surely, after having been kicked across the smoothest ice a rock could stop abruptly. It could return to where it was kicked, or even describe circles.[1] This could happen without ground glass, magnets, or anything else. Is this not possible?'

Here some will reply 'Yes', others 'No'. As before, this is not an experimental issue; it concerns the organization of concepts. The man who says 'No' might continue:

> Once in motion a rock cannot suddenly stop unless something stops it. It cannot return to the kicker's toe unless something brings it back—a magnet, or a jet of air, or invisible threads. It cannot turn circles unless guided by imperceptible grooves in the ice. It would not be a rock, not even a physical body, unless when free of impressed forces it was 'in statu suo quiescendi vel movendi uniformiter in directum'. Anything else is unthinkable.

When others would regard anomalous events as falsifying the law, this person would say 'That only shows the presence of some hidden mechanism.[2] Or else what we took for a rock is not a rock at all.' The first law is less vulnerable to experience for him than for others. He may even regard any event which apparently disconfirms the law statement as itself guaranteeing that (despite appearances) the moving body was not free of impressed forces; or did, in fact, move in a straight line; or was no ordinary physical body. We all reason this way sometimes; physicists observing rocks on ice certainly do so. In the ordinary mechanics of middle-sized bodies a statement of the law of inertia is practically invulnerable. It could hardly be false. Whatever proves a body's motion not to be rectilinear also proves that it is acted on by forces. Thus a form of words, 'If *A* then *B*', at first used so that what it expresses could be false, comes to express what could not be false.[3]

Alcohol boils at 78·3° C. Many people, even alcoholics, do not know this. But most of them know what to look for when asked 'Is that fluid alcoholic?', 'Is there alcohol in that beaker?', 'Is the liquid in the beaker boiling?' and 'What does the thermometer indicate?'. They know how to answer 'What did the thermometer indicate when the alcohol boiled?'. We learn empirically that it always reads 78·3° C.;[1] and so invariant is this that it is virtually part of what we mean by 'alcohol'—at least in physics. A fluid that does not boil at 78·3° C. is not alcohol. That it should be is inconceivable. Similarly, the idea of a rock moving in a circle *proprio motu* over ice makes the physicist's imagination boggle.

In general, to say that something is A (e.g. alcohol) is to remark a characteristic cluster of properties a_1–a_n (e.g. a clear, bright liquid with a unique odour and viscosity). To say that something is B (e.g. boiling) is also to remark a cluster of features b_1–b_n (e.g. an agitated fluid whose surface is broken with bubbles and steam). Put A in circumstances C (where c_1–c_n involves being in a hot beaker containing a thermometer registering 78·3° C.). The result of a few trials of this might be summarized: 'If A is put in C it becomes B.' If shortly after these few trials we find a_1–a_n in circumstances c_1–c_n, but b_1–b_n absent, we might quickly say: 'So it is not really true that any A placed in C becomes B.' If, however, we *never* happen upon a_1–a_n in c_1–c_n where b_1–b_n are absent, then the property 'becoming B in C' may get built into the meaning of 'A'. This is not bound to happen, but it may, and often it does. When it does, the form of words 'A in C is B' becomes a formula permitting us to infer directly, and without possibility of error, from something's being an A in C to the presence of B.[2] At first 'A in C is B' simply summarized a few trials of A in C. The occasional absence of B could have been countenanced, just as we can now countenance a piano with red keys, or a Cambridge winter without rain. B's absence would only have led us to deny that every case of A in C is also B. But when 'b_1–b_n' is put into the meaning of 'A is in C', the absence of B when A is in C is inconceivable. Whatever colour its keys, a piano must be a percussive stringed instrument. A Cambridge winter must include Saint Valentine's day, whatever the humidity. And whatever else alcohol may do, it must boil at 78·3° C.

The laws of physics, of particle physics especially, are used sometimes so that disconfirmatory evidence is a conceptual possibility, and sometimes, as above, so that it is not. This is not the historical point that physical laws begin life as empirical generalizations, but (through repeated confirmations, and good service in theory and calculation) they graduate to being 'functionally *a priori*'. Lenzen and Pap mark this well; Broad concedes it, but insists that the 'cash value' of law statements always rests in their relation to observation; Poincaré demurs, on the grounds that the laws of physics must keep in touch with experience. But the possible orderings of experience are limitless; we force upon the subject-matter of physics the ordering we choose.[1]

These authors regard the shift in a law's logic (meaning, use) as primarily of genetic interest. They agree that at any one stage in the development of physics a law is treated in just one way, as empirical or as 'functionally *a priori*': in 1687 the law of inertia was apparently nothing but an empirical extrapolation; but in 1894 it functioned mostly in an *a priori* way. But this attitude is inadequate. It derives from the belief that a law sentence can at a given time have but one type of use. But the first law sentence can express as many things named 'The Law of Inertia' as there are different uses to which the sentence can be put. Now, as in 1894 and in 1687, law sentences are used sometimes to express contingent propositions, sometimes rules, recommendations, prescriptions, regulations, conventions, sometimes *a priori* propositions (where a falsifying instance is unthinkable or psychologically inconceivable), and sometimes formally analytic statements (whose denials are self-contradictory). Few have appreciated the variety of uses to which law sentences can be put at any one time, indeed even in one experimental report. Consequently, they have supposed that what physicists call 'The Law of Inertia' is a single discrete, isolable proposition. It is in fact a family of statements, definitions and rules, all expressible via different uses of the first law sentence. Philosophers have tendered single-valued answers to a question which differs little from 'What is *the* use of rope?'. Once having decided their answers, they have to deprecate other obvious and, for their points of view, awkward uses of dynamical law sentences.

B

Consider in detail the second law: 'Change of motion is proportional to the motive force impressed and acts in the right line on which the force is impressed.'[1]

Many stress the experiential root of this law statement, as they do with the first. The experience rests in the sensations accompanying muscular exertion when we pull, push and lift.[2] This effort, our experience of which is apparently direct and not further definable, we call 'force'. The direction of a moved body's acceleration is that in which we work our muscles in moving it. So, like acceleration, force is representable in vector notation.

Different amounts of force are required to produce a given accleration in, for instance, a cannon ball and a tennis ball. Conversely, a given amount of force will produce different accelerations in these bodies. However, the direction of acceleration is constant for all bodies—cannon balls and tennis balls alike. Therefore, to each body must be assigned a certain scalar property; let us call it 'the inertial mass m'. The simplest equation embracing all we have so far accounted for is:

$$F = ma = m(dv/dt) = m(d^2s/dt^2).^3$$

Forces derive from many sources, of which muscle power is but one variety.[4] Physics in general is concerned with the nature of these; but mechanics simply takes force as given, whatever their nature. It is concerned only with computing their effects, not their genesis.[5] $F = m(d^2s/dt^2)$ allows essential computations to be made, but within mechanics questions about what 'F' represents are irrelevant.[6]

Nonetheless, '$F = m(d^2s/dt^2)$' has many distinct uses within mechanics. Consider these accounts:

1. F is *defined* as $m(d^2s/dt^2)$. In dynamics that is what 'F' means. It would be self-contradictory to treat 'F' as if it were not strictly replaceable by '$m(d^2s/dt^2)$'. (This is like our earlier examples.)

2. It is psychologically inconceivable that F should be other than $m(d^2s/dt^2)$. A world in which this did not obtain might as a matter of strict logic be possible, but it is not a world of which any consistent idea can be formed. On this equation rests all macrophysical knowledge. Were the world not truly described thus, the system, so useful in dealing

with machines, tides, navigation and the heavens would crash into unthinkable chaos.

3. Perhaps, despite all appearances, $F=m(d^2s/dt^2)$ is false—unable adequately to describe physical events. Perhaps another set of conceptions could be substituted. Nonetheless this would be unsettling. $F=m(d^2s/dt^2)$ facilitates the collection and organization of a mountain of facts and theory. It patterns our ideas of physical events coherently and logically. So the second law, though empirical, cannot be falsifiable in any ordinary way, as are the statements which follow from initial conditions in accordance with this law.

4. $F=m(d^2s/dt^2)$ summarizes a large body of experience, observations, and experiments of mechanical phenomena. It is as liable to upset as any other factual statement. Disconfirmatory evidence may turn up tomorrow. Then we should simply write off $F=m(d^2s/dt^2)$ as false.

5. $F=m(d^2s/dt^2)$ is not a statement at all, hence not true, false, analytic, or synthetic. It asserts nothing. It is either:

(*a*) a rule, or schema, by the use of which one can infer from initial conditions; or

(*b*) a technique for measuring force, or acceleration, or mass; or

(*c*) a principle of instrument contruction—to use such an instrument is to accept $F=m(d^2s/dt^2)$, and no result of an experiment in which this instrument was used could falsify the law; or

(*d*) a convention, one of many ways of construing the phenomena of statics, dynamics, ballistics and astronomy; or

(*e*) '$F=m(d^2s/dt^2)$' demarcates the notation we accept to deal with macrophysical mechanics. Our concern here, (*a*)–(*e*), is not with the truth or falsity of the second law. We are interested only in the utility of $F=m(d^2s/dt^2)$ as a tool for controlling and thinking about dynamical phenomena.

The actual uses of '$F=m(d^2s/dt^2)$' will support each of these accounts.[1] This means not just that among physicists there have been spokesmen for each of these interpretations, but that a particular physicist on a single day in the laboratory may use the sentence '$F=m(d^2s/dt^2)$' in all the ways above, from 1–5, without the slightest inconsistency. Examples of this follow.

Every physics student knows of Atwood's machine. Two unequal masses, m_1 and m_2, are fixed to the ends of a (practically) massless thread, running over a (practically) massless, frictionless pulley. Assign the following arbitrary values to m_1 and m_2:

$$m_1=48 \text{ gm.},$$

$$m_2=50 \text{ gm.}$$

Then,[1]

$$a = 980\frac{50-48}{50+48} = 980\frac{2}{98} = 20 \text{ cm./sec.}^2.$$

This is predicted by the second law.

A well-known physics book follows a similar account with the query: 'Suppose we perform the above experiment and find experimentally a value for a which agrees with the predicted value... *Does it mean that we have proved Newton's second law?*' The author continues, '...this question is absurd, since Newton's second law is a definition and hence incapable of proof...the Atwood machine is essentially a device for measuring the acceleration of gravity g by the determination of a rather than a set-up for the verification of Newton's second law.'[2]

This exemplifies account 1. Physicists do use the second law sentence to express a definition when they need to; they have done so for three centuries. When so used, any statement potentially contradictory to what the sentence expresses may be dismissed as absurd.[3] George Atwood himself found it useful so to use the second law sentence.[4] However, were a statement of the second law nothing but 'a definition and hence incapable of proof', Atwood would have wasted his time in writing his *Treatise*. For his famous machine was invented solely to demonstrate the empirical truth of the law. In the eighteenth century a statement of the second law was regarded universally as a 'substantive statement'; a contingent, universal, descriptive proposition. Atwood remarks: 'The laws of motion...ought not only to be strictly consistent among themselves, but with matter of fact...since any single instance which could be produced of a disagreement or inconsistency would invalidate the whole theory of motion....'[5]

The object of Atwood's neglected *Treatise* was to show that attacks by Bernoulli, Leibniz and Poleni on the law's validity rested on improperly constructed apparatus.[6] He wished to verify it as a substantive statement of fact.[7] With an accurate scale mounted behind m_1, a well-made pendulum, a silk thread of negligible mass and a light pulley (mounted in four friction wheels), Atwood showed[8] that when $m_1 = 48$ gm. and $m_2 = 50$ gm. then the acceleration of m_2 is indeed 20 cm./sec.^{2}. The results were carefully recorded and generalized: they squared with the predictions of the second law. For Atwood this fully confirmed the law.

The point is, if Atwood believed his experiment to verify the statement of the second law, then it must have been thought possible for the machine to have turned up evidence against the law statement. If nothing can falsify a proposition, nothing can verify it either. It was logically possible that m_2 should have accelerated at 5 cm./sec.^{2}, or 50 cm./sec.^{2}. This exemplifies account 4; the second law sentence was used as a contingent universal statement, against which disconfirmatory evidence might weigh at any time. Doubtless this commended itself to Broad when he wrote:

It is certain that the Second Law, as originally stated, was not intended for a definition of force but for a substantial statement about it. Unquestionably the sensational basis of the scientific concept of force is the feelings of strain that we experience when we drag a heavy body along, or throw a stone, or bend a bow.[1]

It is certain also that Newton often puts the sentence to this use.[2] So we have two distinct uses to which physicists have put '$F=m(d^2s/dt^2)$'. They have used in different ways the sentence expressing the law statement: as the result of definitions (account 1), and as an empirical generalization (account 4). Other uses must be considered as well.

Account 2 suggested a use of the sentence which, while expressing what obtains in nature, still seemed inhospitable to any idea of evidence against the law. Indisputably, physicists do use laws in this way, now as in the eighteenth century. Thus Atwood says: '(The) Laws of Motion are assumed as Physical Axioms;... although the mind does not assent to them on intuition, yet as they are of the most obvious and intelligible kind...appear the most proper to be received as principles from which the theory of motion in general may be regularly deduced.'[3] He continues: 'These three physical propositions, having been assumed as principles of motion, reduce the science of mechanics to mathematical certainty, arising not only from the strict coherence of innumerable properties of motion deduced from them *a priori*, but from their agreement with matter of fact.' And then, 'There is no kind of motion but what may be referred to (these) three easy and obvious propositions, the truth of which it is impossible to doubt.'[4] Compare William Whewell, writing in 1834:

The laws of motion...are so closely interwoven with our conceptions of the external world, that we have great difficulty in conceiving them not to exist, or to exist other than they are....If we in our thoughts attempt to divest matter of its powers of resisting and moving, it ceases to be matter, according to our conceptions, and *we can no longer reason upon it with any distinctness.* And yet...the properties of matter...do not obtain by any absolute necessity...there is no contradiction in supposing that a body's motion should naturally diminish.[1]

Physicists often use law sentences as described in account 2. They regard them as empirically true, and yet such that evidence against them is unthinkable.

Philosophers may think these physicists are confused; but the confusion is a difficult one to resist. Certain systems of propositions are empirically true; and therefore the fundamental propositions on which such systems rest must (in some sense) be empirically true as well. However, they are often treated as axioms; they delimit and give definition to the subject-matter to which the system can apply. But nothing describable within the system could refute the laws. Disconfirmatory evidence counts against the system as a whole, not against any of its fundamental parts; it only shows that the system does not hold where it might have held. No part of classical mechanics *per se* enumerates contexts in which it will apply, so no part of the system is proved false when it is discovered not to apply in some context.[2]

Law statements, then, are empirically true, because the system in which they are set is empirically true; but counter-evidence does not disconfirm them. Only in terms of law statements can evidence relevant to the (lower-level) hypotheses of the system be appreciated as confirmatory or disconfirmatory.

Account 4 minimizes this systematic setting of the second law statement. Sometimes it is right and proper to do this. But sometimes the physicist is concerned with the *system* of dynamics, within which nothing disconfirms the laws because they determine those types of phenomena to which the system can apply.

Suppose no alternative systems of concepts were available with which to describe and explain a type of phenomenon; the scientist would then have but one way of thinking about the subject-matter. Nineteenth-century physics provides an example: aberrations in the perihelion of Mercury made Leverrier uncomfortable; but to have scrapped celestial mechanics then would have been to refuse to think about the planets at all. In this sense classical dynamics is empirically true of macrophysical phenomena. (What system could offer a 'more accurate' account of a collision between billiard balls?) Yet the system is true in such a way that the idea of evidence which would falsify its laws often cannot be formed. Account 2 would on some occasions be supported by most physicists, in theory and in practice.[1]

This leaves accounts 5 and 3 to be discussed. Account 5 is familiar enough. When invoked as an 'inference pattern' a statement of the second law is not likely to be called into question by any of the conclusions it warrants. Would $(p.(p \supset q)) \supset q$ be upset by anything inferred in accordance with it? In Atwood's machine, if initial conditions are given as $m_1 = 48$ gm. and $m_2 = 50$ gm. and we wish to infer by way of the second law to the acceleration of m_2, then, if we are actually using the law, the inference pattern itself cannot come under suspicion. It is accepted as a way of reasoning from initial conditions to conclusions.[2]

Similarly, $F = m(d^2s/dt^2)$ can be a 'statement of how force is to be measured for scientific purposes'. Broad advocated this in 1913.[3] But ten years later[4] he dismissed the idea because the measurement of the rate of change of momentum is not the only way to measure force. This strengthens the suggestion that Broad's account is single-valued. Does it follow from the fact that there may be alternative ways of measuring force, that measuring the rate of change of momentum is not *a* way of measuring force? Surely '$F = m(d^2s/dt^2)$' has been used thus. Newton infers from his pendulum experiment that, since different masses have identical constant accelerations towards the earth's centre, a constant force is acting whose magnitude is proportional to the masses of the bodies concerned.[5] This use of the formula predominates in the work of engineers; it inclines some philosophers to regard the second law as *nothing but* a principle of instrument design, or of notation, or of inference.[6] The fundamental formulae of dynamics certainly have such uses, but not to the exclusion of other equally important uses. The same might be said of Broad's emphatic 'single-valued' conclusion: 'The second law, is, therefore, neither a definition nor a statement as to how force is to be measured; but is a substantial proposition, asserting a connexion between two independently measurable sets of facts in nature.'[7]

'$F = m(d^2s/dt^2)$' can sometimes be used to express a definition, sometimes a statement of how force is to be measured, sometimes a substantial proposition (often with disconfirmatory evidence easily conceivable, but sometimes not). What physicists call 'the second law' really consists in everything that can be expressed by way of different uses of this formula.

C

Account 3 has been foreshadowed. To bring out this use further the law of gravitation will serve better than the second law; but before continuing, an apposite quotation from Broad's first book may prepare the way. He writes:

> The true proof of the law [of inertia] is to be found in the explanation that it offers of projectiles' paths and of planetary motion....The nature of the evidence for the Second Law...is in fact precisely the same... viz. that all mechanical processes can be analysed in this particular way.[1]

Broad's answer to our leading question is thus perhaps more than single-valued, and it catches the spirit of the next few pages.

Consider how Newton actually discovered the law of gravitation, remembering what we have learned about Galileo and Kepler. With other seventeenth-century physicists he accepted laws as empirical facts:

1. Each planet moves in an ellipse with the sun in one focus.
2. The radius vector from sun to planet sweeps out equal areas in equal times.

(These two laws were given in 1609.)

3. The squares of the periods of the planets are proportional to the cubes of the mean distances.

(This law was given in 1619.)

Imagine planet P moving elliptically, the sun S being at a focus of the ellipse (see fig. 13). If P's velocity at time t is denoted by v, and if the length of SY (the perpendicular on the tangent at P) is denoted by p, we have

$$pv = h.$$

Produce SY to R, SR being numerically equal to v. If HZ is the perpendicular on the tangent at P from H (the other focus), SR is parallel and proportional to HZ. For the ratio of the lengths we have

$$\frac{SR}{HZ} = \frac{SR.SY}{HZ.SY} = \frac{h}{b^2}$$

(this follows from the property of the ellipse, $SY.HZ=b^2$). If C is the centre of the ellipse, CZ is parallel to SP. We need polar co-ordinates, r denoting SP, and Θ the angle ASP measured from the perihelion A. Turn SR through a (positive) right angle. Then the path R is a hodograph of the motion of P; the velocity of R represents the acceleration of P, and will be perpendicular to SP. This is verified as follows: the velocity of R is parallel to that of Z, and the locus of Z (and Y also) is the auxiliary circle of the ellipse. The velocity of Z is perpendicular to CZ, so the velocity of R is perpendicular to CZ, and thus perpendicular to SP.

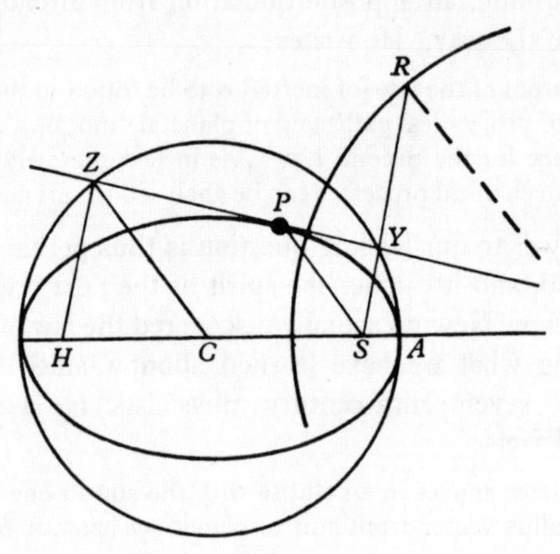

Fig. 13

The acceleration of P is in the direction PS. Its magnitude is measured by the speed of R, which is h/b^2 times the speed of Z (that is, the speed of R is $(h/b^2)\, a\dot{\theta}$). Since $r^2\dot{\theta}=h$, the speed of R is μ/r^2, where $\mu=h^2a/b^2$. Thus *the acceleration of P is in the direction PS and of magnitude μ/r^2*. This is the main result we seek.

Furthermore, the area swept out in unit time by the radius vector is $\frac{1}{2}h$. Therefore the periodic time σ for the elliptic orbit is $2\Delta/h$, where Δ is the area enclosed by the ellipse. Thus

$$\sigma=2\pi ab/h.$$

Now, as we said, $\mu=h^2a/b^2$. Substituting for h in these two equations we get

$$\mu=4\pi^2a^3/\sigma^2.$$

By 3 (p. 105), a^3/σ^2 has the same value for all the planets—that is, the coefficient μ is the same for them all. The acceleration at distance r (from S) is the same for all the planets. Therefore the force on each planet is proportional to its mass.

This reasoning parallels Newton's own; and clearly from here it is a short step to the law of gravitation. The attraction on a planet of mass m is proportional to m/r^2. If there is a universal law of gravitation, it is expected that the masses of any two particles will appear symmetrically in the formula for the attraction, giving the general law

$$\gamma(Mm)/r^2$$

(in the example above μ is γM, where M is the mass of the sun).[1]

It is in terms of this move from Kepler's laws to $F=\gamma(Mm)/r^2$ that account 3 may be developed. For, as Newton here uses the sentence expressing the law, the law statement is clearly empirical: he sets out a property of any pair of particles, punctiform-masses. Alternative states of affairs are possible;[2] the formula is neither necessary nor unfalsifiable. (Indeed, before 1671 Newton *abandoned* the law because he calculated a lunar deflexion of 0·044 in. The observed value was 0·0534.[3] The discrepancy arose from Newton's treatment of a degree of the earth's circumference as 60 miles in length. In 1671 Picard corrected this to 69 miles. This brought about close agreement.)

Nevertheless, $F=\gamma(Mm)/r^2$ did remarkable work in the *Principia*. It could not have just been one more falsifiable, empirical, general statement, like 'all stones in Trinity Great Court contain flint', or 'all flint strikes fire with steel', for the law unified the laws of Kepler and Galileo into a powerful pattern of explanation—one of the most powerful in the history of physics. For Newton the law, as it was first retroduced, did not simply 'cap' a cluster of prior observations: it did not summarize them. Rather it was discovered as that from which the observations would become explicable as a matter of course. Newton was not an actuary who could squeeze a functional relationship out of columns of data; he was an inspired detective who, from a set of apparently disconnected events (a bark, a footprint, a *faux pas*, a stain) concludes 'The gamekeeper did it'. No one less than a Newton, given the laws of Galileo and Kepler, observations of the lunar motion, the tides, and the behaviour of falling bodies, could infer that $F=\gamma(Mm)/r^2$. This law organizes and patterns all these things, and others as well, but nothing incompatible with any of them.

The conceptual situation is not unlike this: novel mathematical theorems are encountered which, besides being individually surprising, do not seem to fit together as a system.

If you accept these particular axioms, such novel theorems, and indeed all mathematics, will (as a matter of course) fall into an intelligible deductive system.

But why should I?—They seem neither clear nor obvious.

Because if you do accept them these particular theorems, and all mathematics, will (as a matter of course) fall into an intelligible deductive system. What could be a better reason?

Similarly, the laws of Galileo and Kepler are discovered, the lunar motion and the tides are studied; but these do not seem to fit together in any systematic way.

If you accept the law of gravitation, the laws of Galileo and Kepler, the lunar motions and the tides will, as a matter of course, be systematically explained and cast into a universal mechanics.

But why should I? The empirical truth of the law is not directly obvious, nor can what it asserts be easily grasped.

Because if you accept it all these things will, as a matter of course, be systematically explained and cast into a universal mechanics. What could be a better reason?

This kind of reasoning gave birth to modern theoretical physics, research within which might be described as observation statements in search of a premiss. Thus: electron beams are deflected in a transverse magnetic field, but electron beams will also diffract when shot through a crystal or a thin metal foil.

If you accept this concept of electron, having properties α, β, γ (e.g. a motion formally analogous to the translation of a wave group, collision behaviour like a classical point-mass, no precisely determinable simultaneous position and velocity), then a comprehensive and systematic explanation of electron deflection, diffraction and of a fundamental uncertainty in microphysical experimentation will follow as a matter of course.

But why should I accept this concept of an electron; since as such it is not even conceivable? 'Wave-group', 'point-mass'? The entity described can be no more than an ingenious mathematical combination of physically distinct parameters.

You should accept it because if you do a comprehensive and systematic explanation of these diverse and apparently incompatible microphysical phenomena will follow as a matter of course. What could be a better reason?[1]

This retroductive procedure, this reasoning back from observations to formulae from which the observation statements and their explanations follow, is fundamental in modern physics. Yet it is least appreciated by philosophers, so often are they attentive to the (indispensable, but sometimes over-estimated) empirical correlations of men like Boyle, Cavendish, Ampère, Coulomb, Faraday, Tyndall, Kelvin and Boys. Philosophers sometimes regard physics as a kind of mathematical photography and its laws as formal pictures of regularities. But the physicist often seeks not a general description of what he observes, but a general pattern of phenomena within which what he observes will appear intelligible. It is thus that observations come to cohere systematically. We ought not to expect the same coherence and intelligibility of the fundamental formulae which so order observations. That they order the observations is their *raison d'être*; to expect the same comprehensibility in these formulae would be like expecting a mechanical explanation of the laws which make mechanical explanation possible. The great unifications of Galileo, Kepler, Newton, Maxwell, Einstein, Bohr, Schrödinger and Heisenberg were pre-eminently discoveries of terse formulae from which explanations of diverse phenomena could be generated as a matter of course; they were not discoveries of undetected regularities. It is this which now drives theorists to search for the root of all of our inverse square laws, dynamical, optical, electrical, and which spurs them on towards a formalism in quantum physics which will not be quite so productive of procedures which are, mathematically, quite *ad hoc*. It drives Heisenberg to dream of 'a single formula from which will follow the properties of matter in general'. All this is behind accounts 2 and 3.

The law of gravitation can provide a conceptual Gestalt; for Newton it was a new pattern for mechanical thinking, and could not have been falsifiable for him in the way that the statements patterned by the law were falsifiable. It made the laws of Kepler cohere for Newton as they did not cohere for Kepler himself, although Kepler had everything necessary for the geometrical retroduction above. That the equation could do this puts it into a special class of uses of law sentences. Here $F=\gamma(Mm)/r^2$ is clearly empirical, but its use is different from that of any gross generalization about all observable members of a certain class, and different again from a rule of inference, a definition, or a principle of measurement. And if its use is different its meaning is different, and if its meaning is different its logic is different.

One might risk saying that the law of gravitation is sometimes regarded as *a priori* because it is synthetic—that is, it synthesizes a diversity of observation statements. From this use it is a short step to account 2. Here $F=\gamma(Mm)/r^2$ is an empirical assertion about the relations between particles in the universe, but it is psychologically inconceivable that evidence against it should turn up. We may move next to account 5, where the law provides a pattern of inference, a way of reasoning from initial conditions (such as the present position and disposition of Mars) to predictions (such as the position and disposition of Mars on St Valentine's day). Since Newton, this has been a central use of '$F=\gamma(Mm)/r^2$'.

With the successes of classical mechanics in the nineteenth century, the law often came to be construed according to account 1, as a definition.[1] Contrast this with the uses of the law statement, in the last three centuries, as a straightforward empirical hypothesis —as in account 4. We saw how Newton once abandoned his gravitational theory as an hypothesis which did not square with the facts. The eighteenth century found scientists, Clairault and Buffon, for instance, disputing whether or not $F=\gamma(Mm)/r^2$ was the necessary and inevitable law for any gravitational force.[2] Newton found that the line of the moon's apsides sweeps through the heavens with a velocity twice as great as the law seems at first to give;[3] this was the only failure of the theory. It was discovered later that apparently insignificant formal residues, dismissed in the course of the lengthy calculation, were cumulatively important. Until then, however, the law seemed at fault. Clairault tried to help; he introduced a small additional force varying inversely as the square of the fourth power. Buffon countered that the force *could not* vary according to any law other than the inverse square; in his opinion so many facts supported the law that this single discrepancy had to be explained away. Clairault objected that the law of gravitation does not obtain exactly, noting that in cohesion, capillary attraction and other cases, forces vary according to laws other than the inverse square. Into this controversy came the experimentalists, who were all eager to exploit the method (indicated by Newton) by which γ might be determined if only M and m were known. When M is of astronomical dimensions the problem is unmanageable; but in 1740 Bouguer showed the mass of mountains to be measurable, and in 1774–6 Maskelyne estimated the mass of Mount Schiehallion. (He observed the deflexion of a plumb line on opposite sides of the mountain.)[1]

Then Cavendish, in 1798, using a delicate torsion balance, discovered an observable attraction between a heavy and a light metal ball.[2] As he says: 'These experiments were sufficient to shew, that the attraction of the weights on the balls is very sensible, and are also sufficiently regular to determine the quantity of this attraction....'[3] Cavendish considers the objection 'that it is uncertain whether, in these small distances, the force of gravity follows exactly the same law as in greater distances'. He concludes that 'There is no reason, however, to think that any irregularity of this kind takes place'.[4] One possible result of this experiment was that *no* deflexion should have been observed. That it was observed (confirming that there exists an attraction between all physical bodies), rests on the possibility that Cavendish's experiment might have turned out to be wholly disconfirmatory. Had this been so, $F=\gamma(Mm)/r^2$, in this context, would simply have been false.

The experiment of C. V. Boys in 1895 was more finely designed.[5] Note the recent date; the gravitation law was still regarded as an empirical hypothesis late in the history of dynamics (well after the axiomatic treatment of Hertz). This counters the view of Pap and Lenzen, who regard the move from empirical hypothesis to functionally *a priori* principle as the evolutionary pattern in physics. The ideas of Michell and Cavendish were modified by Boys, whose apparatus had an accuracy of 1 in 10,000.[6] He discovered that the force of attraction between a large freely swinging lead ball M, and a tiny freely swinging gold ball m was $6 \cdot 6576 \times 10^{-8}$ dynes. (Incidentally this showed the density of the earth to be $5 \cdot 5270$ times that of water.)

Thus for over two centuries '$F=\gamma(Mm)/r^2$' had a use as an empirical statement, potentially falsifiable. From Newton to Boys (1895) gravitational attraction was something to be established within experimental mechanics. Yet from Newton to Hertz (1894), '$F = \gamma(Mm)/r^2$' had important uses as a principle of inference within axiomatic mechanics, was often set out in texts as a definition, and was sometimes a principle of instrument construction. Finally, it was invoked in the way we have alluded to earlier, sometimes as an empirical truth whose contradictory was consistent but psychologically inconceivable, and sometimes as an empirical truth unifying other bodies of information, but whose contradictory was both consistent and conceivable. These last reflect the success of classical mechanics in shaping and patterning our physical ideas. The scope of that system has diminished somewhat in this century, but, now as before, law sentences and formulae can be used in a variety of ways in conducting and reporting experiments. This seems contradictory only when the laws of a system are thought to be single-valued in their use.

D

Single-valued accounts of the nature of dynamical laws are all supported to some degree by the practices of physics; consequently each special account fails to be supported completely and exclusively. 'The second law of motion', 'the law of gravitation': these have been construed as the names of discrete propositions. But in physics they are umbrella-titles; they cover everything that '$F = d^2s/dt^2$' or '$F = \gamma(Mm)/r^2$' can be used to express—definitions, *a priori* statements, heuristic principles, empirical hypotheses, rules of inference, etc. One experimental report may employ a given formula to express each of these in turn; and so raises questions.

Perhaps this only shows that the term 'laws of dynamics' has a different force for philosophers and physicists. It may even be urged that the account is tainted with 'psychologism': why should every use by physicists of '$F = \gamma(Mm)/r^2$' be relevant to the philosophical question 'What is the logical status of the law of gravitation?'. Is every use of '$F = \gamma(Mm)/r^2$' in physics a correct use? A physicist's uses of formulae are not germane to determining what the laws of dynamics are; nor is what he *thinks* about laws—any more than is his behaviour when drunk, doped, or distracted. Nor are the processes involved in discovering a law pertinent to assessing the logic of that law.

The objection rests on an unacceptable realism. What concept have we of the laws of dynamics, other than that gained by studying the uses to which dynamical law formulae are put in physics? What is the form of our concept other than that expressed in the *Principia*? Short of becoming Newtons ourselves and rethinking all of physics *ab initio*, how else could we grasp the meaning of mechanical laws? Philosophers can but record, compare and analyse the positions of mechanical law statements in the concept-system that is classical mechanics. There are no workable concepts of the laws of dynamics other than what these sentences can be used to express in mechanics.

Philosophy of physics is thus unique: most specialized disciplines become pure philosophy when pushed to their fundamentals, but physics becomes 'natural philosophy'. When pursued to its foundations it is conceptual analysis, criticism of criteria and revision of methods and ideas. The most eminent inquirers in this field have not been academic philosophers but men like Kepler, Galileo, Newton, Maxwell, Einstein, Bohr, Schrödinger and Heisenberg. They discovered what the laws of mechanics were, and what was their status within physics. The philosopher of science, unless he is also one of these natural philosophers, can only try to understand and elucidate what the laws do in the actual solving of physical problems and in the thinking of physicists; which is just to trace how mechanical laws are used in patterning otherwise perplexing phenomena. To go further—to make philosophical recommendations about the *real* nature of the laws of dynamics or about how law formulae *ought* to be used—is exactly what the theoretical physicist is trained for. The philosopher of physics who ventures into this territory must expect to be judged by standards unknown in the British Academy.

'But surely not all the uses of dynamical law formulae are *equally* relevant to understanding the laws of dynamics?' What could be more relevant than some proof that dynamical laws are all *a priori* or that they are all contingent? But even if there were such a proof it would only be a way of remarking certain typical uses to which law sentences can be put. To use a formula so that it expresses a statement whose negation is self-contradictory is to hold that statement to be analytic; evidence offered to support the negation would be dismissed without a hearing. For the law statement to be called 'contingent' is only for a certain typical use of the formula expressing it to be specified; with respect to this use, evidence against the statement is clearly possible.

Is this psychologism too? Can one understand the logic of statements without understanding how people use the sentences which express them? How? Can we say 'It is irrelevant how anyone or everyone uses dynamical law formulae or what anyone or everyone thinks about mechanical laws. The question is whether or not the laws *themselves* are necessary or contingent'? What are the laws of dynamics *themselves*? What are they other than what dynamicists express by the use of law formulae? What is the concept intended in the question? *Is* there a philosophical alternative to the physical concept of the law of force or of gravitation? If so, by what criteria is it to be judged? Can the logic of a proposition be grasped in any way other than by learning how the sentence expressing it is used on particular occasions by particular people? The logical status of the laws of dynamics is revealed when it is shown how, for instance, '$F = d^2s/dt^2$' and '$F = \gamma(Mm)/r^2$' are used by physicists doing physics. But these uses are not limited to those describable by 'necessary', or 'contingent'. All the uses of law formulae must be equally relevant to understanding the laws of dynamics. Accounts 1 and 4 are not specially to be preferred.

Once embedded in a theory, a law which was originally contingent joins a family of other assertions, all of which may be expressed by the same law sentence. Some members of this family can only be described as *a priori*. Further observations may be made about this.

A law sentence expresses an *a priori* proposition when its user maintains it in the face of all experience. He may do this because he regards the idea of potentially falsifying evidence as (1) impossible on logical grounds, i.e. self-contradictory, or (2) impossible on physical grounds, e.g. inconsistent with a conservation principle, or (3) consistent, but psychologically inconceivable. (3) could be dismissed as logically irrelevant, but one might risk the heresy of saying that this is debatable. If not one physicist has even a workable conception of x (e.g. levitation in terrestrial space), will this not affect the use and hence the meaning of terms and formulae? The logical point is that some laws *are* maintained in the face of all experience. (1), (2) and (3) are the reasons why; all three have been invoked to explain our inability to conceive a fast particle at a geometrical point. A man might hold to a law against all counter-experience because, for instance, a Pope or a Commissar instructed him to do so. We deplore his reasons, but the logical issue remains unaffected: he uses the law sentence concerned in an *a priori* manner.

Dynamical law sentences are used in many contexts to express what is non-contingent. Just as a collar stud used to replace a lost pawn *is* a chess token, so a proposition which has a contradictory which is (1) logically impossible, or (2) physically impossible, or (3) psychologically impossible will be *a priori*. When a genuine disconfirmatory instance does appear, rather than that their universality should be qualified, dynamical laws are usually 'saved'. The law's universality is retained, but it is made inapplicable to the recalcitrant instance. This may be strictly equivalent to restricting the law's universality; but in physics the two procedures would always be distinct. Suppose that '$F = d^2s/dt^2$' is used so that its employer maintains it in the face of all possible experience. Does this mean that the law is not an empirical hypothesis? No: account 2 considered the law as an empirical hypothesis, whose denial was not self-contradictory. Nonetheless the law statement would have been maintained whatever happened, since any alternative would have been psychologically inconceivable—conceptually untenable. In a different way this is also true of account 3.

Suppose that '$F = d^2s/dt^2$' were used so that what it expressed was not *in principle* subject to disconfirmation. (It is physical principles that are being considered.) Is the denial of the law so expressed self-contradictory? 'The second law of motion is not, as a matter of principle, subject to empirical disconfirmation.' This is a necessary condition for the proposition 'the second law is analytic'; but is it also a sufficient condition? Or are there uses of '$F = d^2s/dt^2$' in which what is expressed, though not subject to empirical disconfirmation, is nonetheless not analytic (its denial not self-contradictory)? Is there a sense of *a priori* which is different from *analytic* and also from *functionally a priori*?

Today we have perspective on the second law of motion: perhaps too much to be able to feel its force on those eighteenth and nineteenth-century physicists who regarded it as a supreme empirical truth, yet in principle above disconfirmation. Consider some statements which today control our physical thinking in comparable ways:

1. It is impossible for an engine to deliver more work at one place than is put into it at another.[1]

2. A perpetual-motion machine of the second type (thermodynamical) is impossible.[2]

3. Nothing travels faster than light.[3]

4. Nothing whose motion is properly described in terms of $\Delta\psi + 8\pi^2 m/h^2(E-U)\psi = 0$ can have simultaneous position and momentum co-ordinates defined any more precisely than is given in the relations $\Delta p . \Delta v \cong h/m$.

These are for us what the force law must have been for earlier physicists. To deny what they assert would not be self-contradictory, but to say that therefore their denials are only psychologically inconceivable would strike any physicist as frivolous. 'Today is Monday and yesterday was Friday'; 'Two dozen makes 25'. These are absurd, false *a priori*. They could not possibly be true, no matter what happens. There are plenty of *a priori* statements like these which are not purely formal—that is, which do not arise strictly from certain uses of 'and', 'not', 'another', 'either–or', 'all', etc. For example: 'Your first cousin once removed cannot be your parent's cousin's child', and 'The Stars and Stripes cannot have more stripes than stars'. The necessity of these cannot be grasped simply from knowing the use of logical operators. We must appreciate also the uses to which 'cousin', 'parents', 'Stars and Stripes' are actually put.

'In his doctoral thesis he successfully designs a perpetual-motion machine'; 'This entity travels faster than light.' To grasp the absurdity here we must do more than reduce the assertions to some 'Carnapian' symbolism. We must appreciate how differently we would determine the truth or falsity of 'This machine is more efficient than that', 'This is the most efficient machine ever built', *and* 'This machine is perfectly efficient'. We must understand the differences between 'This man runs faster than that one', 'This train travels faster than that van', 'This bullet travels faster than that aeroplane', 'Light travels faster than this bullet', *and* 'This entity travels faster than light'.

The differences are not like the differences between 'This is the Stars and Stripes' and 'This is the Jolly Roger'. The truth-values of these last are determined by procedures of the same type; but the procedure for determining whether or not something travels faster than light is different in principle from those involved in determining whether one man runs faster than another, or whether trains travel faster than vans, aeroplanes, or bullets. No physicist would dream of trying to settle experimentally someone's claim to have constructed a *perpetuum mobile*.

'No machine is perfectly efficient.'

'It is inconceivable that any machine should be perfectly efficient.'

'"This machine is a *perpetuum mobile*" could not but have been false.'

Compare these with:

'What he said { must have been false / could not but have been false / could not have been true } since he said "It is Monday today and yesterday was Friday".' There is a difference between the simple logical moves which show that his statement could not but have been false, when what he said was 'Today is Monday and yesterday was Friday', and the complex non-logical moves which show that a man's statement could not but have been false when what he said was 'He has designed a *perpetuum mobile*', 'He has discovered a particle which moves faster than light', or 'He has precisely determined the instantaneous velocity at a point of a high-energy elementary particle'.

Yet all of these decisions turn on matters of principle. When statements like the last three are used by physicists so that as a matter of principle no procedure would show them to be true, then we will hear it said of them that they could not but have been false—no matter what had happened. The physicist himself would quickly remark that these statements are not false like any gross contingent statement which *might* have been true but is false. Nor are they false like a false mathematical or logical statement, or like one which simply fails to obey prior definitions. Neither will the physicist agree that these statements are false as are 'I am not now reading these words' and 'The earth is flat'—i.e. psychologically inconceivable. Perpetual motion machines and velocities greater than light are not just psychologically inconceivable: the physicist will say that they are *impossible in principle*.

A final remark will both bring us back to earlier considerations and prepare us for the concluding chapter. To say that when '$F=d^2s/dt^2$' is used in several different ways, different propositions are being expressed, is true enough. Yet it is an important fact that everything expressed by the uses of '$F=d^2s/dt^2$' holds together in a most intimate way. In some sense you and I do say the same thing when we say '$F=d^2s/dt^2$' even though we may express quite different propositions. In a particular dynamical problem, such as the determination of the noon position of Mars on St Valentine's day 1960, our paper calculations will be identical, our predictions indistinguishable and our explanations very similar. So, though there is a sense in which we are proceeding differently, there is also a sense in which we are proceeding in the same way (a situation which will by now be familiar to us). You and I are thinking the same things, but differently; the patterns of our thoughts differ. Imagine Mach and Hertz working out this planetary problem: as they were considerable physicists, both would be likely to give a correct answer; yet there would be something different about the quality of their reflexions. Mach construed dynamical laws as summary descriptions of sense observations, while for Hertz laws were highly abstract and conventional axioms whose role was not to describe the subject-matter but to determine it. The difference is not about what the facts are, but it may very well be about how the facts hang together. Even this difference would not seem to matter much here, since Mach and Hertz would get the same answers to their problems. The real difference, however, only arises at this point: for though they get the same answer to the problem, the difference in their conceptual organization guarantees that in their future research they will not continue to have the same problems. Classical mechanics is no longer a research science; *its* problems can be dealt with in almanac fashion. This is why Mach and Hertz get the same answer to their problem. The important difference in conceptual organization, which it has been our aim to illuminate, shows only in 'frontier' thinking—where the direction of new inquiry has regularly to be redetermined. Kepler and Tycho might not have got the same answer to the problem. Beeckman and Descartes did not get the same answer to their problem, though the formulae seemed to be the same for both. Boyle and Newton never even had the same problems: nor did Faraday and Maxwell. Nor did Schrödinger and Born, Einstein and Dirac, Bethe and Heisenberg.

CHAPTER VI

ELEMENTARY PARTICLE PHYSICS

There is no inductive method which could lead to the fundamental concepts of physics...in error are those theorists who believe that theory comes inductively from experience. EINSTEIN[1]

Elementary particle physics is a typical research science. It seems anomalous only when forced into a misleading contrast with classical text-book physics. One must compare the conceptual perplexities of contemporary physicists with those of Galileo, Kepler, Descartes and Newton when they were *creating* physics. A Galileo grappling with acceleration, or a Kepler considering a non-circular planetary orbit, or a Newton reflecting on the particulate nature of matter and light—these do not differ essentially from cases of a Rutherford entertaining 'Saturnian' atoms, or a Compton proposing a granular structure for light, or a Dirac suggesting a positive electron, or a Yukawa wrestling with the idea of a 'meson'. This is frontier physics, natural philosophy. It is analysis of the concept of matter; a search for conceptual order amongst puzzling data.

The similarity of explanation from Thales to today's physicists is striking. We shall in the present chapter consider this in relation to some themes in the pattern of microphysical thinking: A, the nature of atomic explanation; B, the wave-particle duality; C, individuality in microphysics; D, the logical status of the uncertainty relations; E, the correspondence principle, and F, the ψ function and its significance.

A

Physicists advise us not to picture atomic particles. Fermi warned that the search for a picturable electron would lead to confusion. This can be puzzling; for how can one discover and interfere with unpicturable, unvisualizable objects? With what instruments can they be forced into the open? How can such entities be conceived at all? Well, atomic particles *must* lack certain properties; electrons could not be other than unpicturable. The impossibility of visualizing ultimate matter is an essential feature of atomic explanation.

Suppose you asked for an explanation of the properties of chlorine gas—its green colour and memorable odour. Would this satisfy you?—'The peculiar colour and odour of chlorine derive from this: the gas is composed of many tiny units, each one of which has the colour and the odour in question.'[1]

Would this be adequate? Many physicists would not think it an explanation at all. Newton accused those who explained cohesion between bodies by inventing hooked atoms of 'begging the question'.[2] Seeber denied a brick-like structure to crystals; this would require investing the bricks with just those properties of crystals which require explanation. 'This does not solve the problem but only pushes it one step farther back.'[3] Similarly Clerk Maxwell,[4] von Laue,[5] Dirac and Heisenberg,[6] all of whom have noted and rejected this type of 'explanation'. It does not answer questions about material properties, it only postpones them. What requires explanation cannot itself figure in the explanation. Would we be satisfied if the sleep-inducing qualities of opium were explained by reference to the soporific properties of the opium molecule?[7] A soporific property *is* a sleep-inducing quality. The explanation has been deferred.

One might object: the dynamical behaviour of a billiard ball can be explained by the similar behaviour of another ball which has just struck it. One could explain why a cloud moves by referring to the motions of its constituent molecules (whose group motion *is* the cloud's motion). It also explains the redness of blood to say that blood is made up of red particles.

True. But one cannot explain why any given thing is red by saying that *all* red things contain red particles; nor could one explain why any thing moves by noting that any moving thing contains moving particles. In general, though each member of a class of events may be explained by other members, the *totality* of the class cannot be explained by any member of the class. The totality of movement cannot be explained by anything which moves. The totality of red things cannot be explained by anything which is red. All picturable properties of objects cannot be explained by reference to anything which itself possesses those properties.

The history of atomism illustrates this. Greek natural philosophers[8] sought to explain the diversity of physical properties. Thales, Empedocles, and Democritus agreed that the myriad colours, odours, tastes and textures of things were not each one final and irreducible, but could be analysed further; they were the manifestations of something more fundamental.[1] Most nominations for this 'more fundamental something' failed because they possessed the properties to be explained. Water could not be just a liquid, nor a vapour (fog), nor just solid (ice). Were it one of these —liquid, say—how could reference to it explain the solid and vaporous things we observe? But if 'water' named a trinity of types of matter, a 'liquid-vapour-solid', explanation of material properties in terms of it would be complex and mysterious. Were these properties abandoned, however, why should the fundamental substance be called 'water' at all? The concepts of earth, water, air and fire constitute Empedocles' attempt to meet this logical difficulty; by the mixture of these ideal elements he explained all properties of objects. But this was inelegant and uneconomical, and it left the ideas of solid, liquid, vapour and heat themselves unexplained.

Democritus saw that if this fundamental something was to explain the properties of objects, it could not itself possess those properties. Earth, water, air and fire did possess them. His atoms therefore lacked all properties (save geometrical and dynamical ones); they were identical and purged of 'secondary qualities'. 'A thing merely appears to have colour; it merely appears to be sweet or bitter. Only atoms and empty space have a real existence.'[2] This already renders the atom unpicturable—can a colourless atom be pictured? (Windows and spectacles can be pictured because at certain angles they are not transparent.)[3] If the colours of objects are to be explained by atoms, then atoms cannot be coloured, nor pictured. The request to explain the properties of chlorine was not a question about a local phenomenon; this bottle of gas with these properties. A general account of the properties of chlorine was sought which would show how it affects us as it does. Simply to endow atoms of the gas with these same optical and chemical properties is to refuse to supply that theoretical account.

What is it to supply a theory? It is to offer an intelligible, systematic, conceptual pattern for the observed data. The value of this pattern lies in its capacity to unite phenomena which, without the theory, are either surprising, anomalous, or wholly unnoticed. Democritus' atomic theory avoids investing atoms with those secondary properties requiring explanation. It provides a pattern of concepts whereby the properties the atom *does* possess—position, shape, motion—can, as a matter of course, account for the other 'secondary' properties of objects. The price paid for this intellectual gain is unpicturability.

Atomic explanation did not change; scholars remained unable to visualize atoms, just as Democritus' contemporaries had been.[1] As the theory gained support in chemistry and physics, however, scientists came to regard atoms as familiar things.[2] When speaking strictly[3] they renounced the picturable atom; but why speak so strictly? The geometer never denies himself the use of drawn lines: lines *should* be one-dimensional, but proofs and constructions cannot be carried out with one-dimensional lines. Similarly, physicists could think about atoms only by visualizing them. Why not? It helped to secure explanations. Thus the almost invisible diagrams of geometry crept into physical thinking about atoms. Rutherford was thinking on these lines in 1911 when he accepted Nagaoka's idea 'of a "Saturnian" atom which...consist[s] of a central attracting mass surrounded by rings of rotating electrons'.[4] Atoms should have been as unpicturable as the entities of geometry, but no physicist chose so to paralyse his thinking. Indeed, atoms became models of geometrical and dynamical behaviour;[5] and this made them eminently picturable. Why should colours and lines be more than a practical necessity? Like ideal circles, the classical atom was just the limit of a series of sketches of increasing fineness.[6]

Even this expedient no longer serves the imagination. Atomic explanation always ruled out secondary qualities; now modern atomic explanation denies its fundamental units any direct correspondence with the primary qualities, the traditional dimensions, positions, and dynamical properties. In classical physics kinematical studies precede dynamical ones; in quantum physics this division and order is not feasible. Primary qualities were fundamental to the statical-kinematical conceptions which classical particle theory built into a Euclidean space; dynamical properties of bodies were ancillary to these. Now an atomic particle's statical-kinematical properties are determined by its dynamical properties: quantum dynamics is the prior discipline. The basic concept of microphysics is *interaction*. The Democritean-Newtonian-Daltonian atom cannot explain what has been observed in this century. Its postulated properties—impenetrability, homogeneity, sphericity—no longer pattern and integrate our data;[1] to account for all the facts the atom must be a complex system of more fundamental entities.[2] Electrons, protons, neutrons, positrons, mesons, anti-protons, anti-neutrons and γ-ray photons have been detected; others are likely too if certain 'gaps' in our experiments are to be explicable. But these cannot be the point-particles of classical natural philosophy.[3]

The properties of particles are discovered and (in a way) determined by the physicist. Phenomena are observed which are surprising and require explanation. The observations may be of the tracks left in a cloud chamber or in a photographic emulsion, or the scintillations excited when particles strike certain sensitive screens, or one of a number of their other indirect effects.[4] The theoretician seeks concepts from which he can generate explanations of the phenomena. From the properties he ascribes to atomic entities he hopes to be able to infer to what has been encountered in the laboratory: he aspires to fix the data in an intelligible conceptual pattern. When this is achieved he will know what properties fundamental entities do have; and he will have learned this by retroduction.

For example, electrons 'veer away' from negatively charged matter; they must therefore be like particles. But electron beams diffract like beams of light, and therefore they must be like waves too.[5] The physicist fashions the electron concept so as to make possible inferences both to its particle and to its wave behaviour, and a conception so fashioned is unavoidably unpicturable. Observations multiply; properties are pushed back into the concept of 'electron', properties from which each new explanation follows as a matter of course. Unless this leads to unsound inferences, the theory which depends on the particle having these properties will be felt to explain the observations. Indeed at this point, one could have no reason to doubt the real existence of the properties; intelligibility would demand them of these sub-atomic entities.[6]

The result is radical unpicturability. If microphysical explanation is even to begin, it must presuppose theoretical entities endowed with just such a delicate and non-classical cluster of properties.[7] In general, if A, B and C, can be explained only by assuming some other phenomenon to have properties α, β and γ, then this is a good reason for taking this other phenomenon to possess α, β and γ.[1] In macrophysics any such hypothesis is tested by looking at the other phenomenon to see if it has α, β and γ. With elementary particles, however. we cannot simply look. All we have to go on are the large-scale phenomena *A*, *B* and *C* (ionization tracks, bubble-trails, scintillations, etc.) and perhaps future phenomena *D*, *E* and *F*. Hence one must suppose that the particles actually have the 'explanatory' properties in question, α, β, γ, and see if, by mathematical manipulation of these, we can infer to further theoretical properties δ, ϵ and ζ, which might explain the further phenomena *D*, *E* and *F*.

The cluster of properties α, β and γ may constitute an unpicturable conceptual entity to begin with. As new properties δ, ϵ and ζ are 'worked into' our idea of the particle,[2] the unpicturability can become profound. This does not matter: there can be no atomic particles which we may fail to recognize because we failed to form an identification picture of them in advance. The whole story about fundamental particles is that they show themselves to have just those properties they must have in order to explain the larger-scale phenomena which require explanation.

Thus, discovering the properties of elementary particles consists in a logical situation which is in principle like that in which Democritus found himself. Unless they are thought to have certain abstract properties they cannot explain the phenomena they were invoked to explain; they cannot resolve retroductive inquiries. Professor Fermi illustrates this: 'The existence of the neutrino has been suggested...as an alternative to the apparent lack of conservation of energy in beta disintegrations. It is neutral. Its mass appears to be either zero or extremely small....Its spin is believed to be $\frac{1}{2}$; its magnetic moment either zero or very small....'[3] Our concepts of the properties of the neutrino are determined by there being gross phenomena *A*, *B*, *C*, which defy explanation unless an entity exists having the properties α, β and γ—just those which the neutrino has. The neutrino idea, like those of other atomic particles, is a retroductive conceptual construction out of what we observe in the large; the principles which guarantee the neutrino's existence of electrons, α-particles, and even atoms. This does not make the subject-matter of atomic physics less real. Elementary particles are not logical fictions, or mathematically divined hypotheses spirited from nowhere, to serve as bases of deductions; nor does knowledge of elementary particles consist only in a summary description of what we learn directly through large-scale observation. What we must realize, however, is that knowledge of this portion of the world is derived by means more complex than any such philosophically easy accounts suggest.

Again, the situation is as follows. This surprising phenomenon is observed: we expect the energy released by homogeneous radioactive substances to depend solely on the initial and end stages of the nucleus (hence all α-rays of a homogeneous substance have the same range, i.e. the same energy). But β-particles are emitted with all possible energies (Chadwick). This contradicts the principle of conservation of energy.

Accept the hypothesis (of Pauli): with every β-particle another particle also leaves the nucleus, carrying the difference in energy. If this particle is construed (following Fermi) as having the properties: velocity c, hence mass $=0$ and in no case greater than 1/500th an electron mass, charge neutral, magnetic moment $=0$ (or very small), then the continuous spectrum of the β-ray will be explicable as a matter of course, and the energy principle still holds.

Yes, but why accept this concept [of the *neutrino*]? It cannot be observed in the Wilson chamber, nor has it ever been *directly* detected by any other means. Besides, such a particle seems unlikely and unsettling. So why accept the neutrino?

Because if you do, the continuous β-ray spectrum will be explained as a matter of course, and the energy principle will remain intact. What could be a better reason?[1]

The formation of the neutrino concept provides a paradigm example of how observation and theory, physics and mathematics, have been laced together in physical explanation. Mathematical techniques more subtle and powerful than the geometry of Kepler, Galileo, Beeckman, Descartes and Newton are vital to today's physical thinking. Only these techniques can organize into a system of explanation the chaotically diverse properties which fundamental particles must have if observed phenomena are to be explained. As Heisenberg puts it: 'The totality of Schrödinger's differential equations corresponds to the totality of all possible states of atoms and chemical compounds.' He even dreams of

'...a single equation from which will follow the properties of matter in general'.[1]

As concerns mental pictures, the present situation in fundamental physics could not have been different. This is unpicturability-in-principle: to picture particles is to rob oneself of what is needed to explain ordinary physical objects.[2] Though intrinsically unpicturable and unimaginable, these mathematically described particles can explain matter in the most powerful manner known to physics. Indeed, only when the quest for picturability ended was the essence of explanation within all natural philosophy laid bare.

B

Before we explore the idea-pattern of elementary particle physics, an interlude may help to disclose the situation from which it springs.

Imagine a mid-nineteenth-century physicist teaching mechanics to his freshmen. Unfortunately, he is behind schedule, so he decides to double up his demonstrations, putting them 'end to end'; this works very well. He comes to Galileo's spheres on the inclined plane, and 'doubles up' the experiment by augmenting it as follows. Instead of terminating their descent at the base of the plane, the spheres are allowed to drop. They enter a maze, a kind of Galton board, through whose pins and baffles they scatter. Each sphere finally falls into one of a battery of tube-like receptacles (see fig. 14.) Naturally, the professor expects the balls to collect themselves as shown in fig. 15. Had they done so, freely falling bodies and statistical distributions could have been discussed, all in one hour. But his plans are spectacularly foiled. Inexplicably, the spheres distribute themselves in the way shown in fig. 16: an unlikely distribution, unprecedented in classical particle theory.

This *gedankenexperiment* leads to some improbable things. Why do we think them improbable? Our hypothetical professor could draw but one conclusion in accordance with the knowledge of his day: the distribution constituted an interference pattern, the sign of a periodic, wave-like process. Thus in addition to their obvious particle-like properties, the spheres show wave behaviour too; the upper half of the experiment is definitive of particle behaviour, the lower half of wave behaviour. To describe these findings adequately, the professor would have to combine in one language the only notations appropriate to this experiment, a particle notation and a wave notation. His *concept* of what these spheres are would have to be modified.[1]

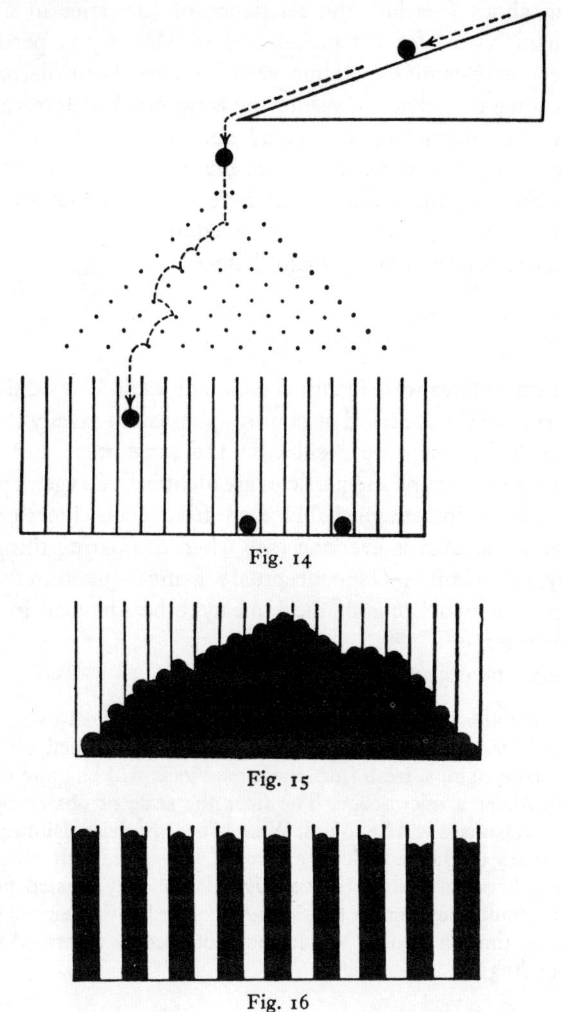

Fig. 14

Fig. 15

Fig. 16

This example parallels experimental discoveries made in the last thirty years. Distributions as classically improbable as that set out above have been encountered with elementary particles; and the conceptual resistance was at least as great. It is the same resistance that young Kepler felt towards a non-circular Martian orbit, the resistance the Galileo of 1604 felt towards a time parameter when thinking about free fall, the resistance of Leverrier to a 'non-Newtonian' cause for the precessions of Mercury at perihelion. Despite preconceptions, atomic particles have been disclosed to possess wave properties: consider the conceptual pattern required to make this disturbing fact stand out. The whole of classical particle theory and wave theory are involved in our hypothetical professor's recognition that he has here a phenomenon for which a combined wave-particle notation is required.

Let us return now to our main themes.

C

Democritus' atoms were identical each with each. They differed in no respect: all the spherical ones were spherical in exactly the same way; the cubic ones were cubic in the same way; and so on. Similarly all *our* atoms and particles are identical. Oxygen 17 atoms are absolutely indistinguishable each from each, identical in a stronger sense than is ever the case when comparing things that are very, very similar. *Our* conceptual scheme—quantum theory—requires that two atoms of the same type be identical in a very strict manner.

It might be objected:

> No two things are ever perfectly identical. Identical twins can be remarkably similar, but they can always be distinguished ultimately. Two postage stamps, fresh from the same block, will be quite different in detail under a microscope. The finer the scale of observation, the more discrepancies will be found. What is the physicist claiming? That two particles of the same kind are *completely* alike, with no possible difference between them whatever? Even were they created perfectly identical, could they remain thus? They 'collide' with their neighbours millions of times a second. Would they not become deformed with all this pounding?

Macrophysical objects can get into different states and still remain the same object. But an atom or an elementary particle (as conceived in quantum theory) has a strictly limited number of possible states. If we inquire about states differing even slightly from the normal one, it often turns out that there are none. The collisions an atom usually has are not violent, so it simply stays in its normal state. At high temperatures collision is more violent, and an atom may indeed be in a different state after such a collision. But not for long: it quickly gets rid of excess energy with a flash of radiation; it then 'falls' back to its normal state of lowest energy. So the atom is unlike any macrophysical object in at least two ways: (*a*) it is a *perfectly* elastic body in collisions that are not too violent, and (*b*) when it is 'deformed' in violent collisions it returns to its former state with no difference whatever. If two atoms are created identical, then according to the conceptual pattern of quantum theory they will stay identical. How can we know that they were identical to begin with? How could we detect infinitesimal differences between, say, two atoms of oxygen 16?

When Rutherford (following Nagaoka) conceived of the atom as a miniature solar system—electrons circling the nucleus as planets circle the sun—some philosophers suggested that electrons might really be planets with mountains, oceans and even living creatures. Perhaps these in turn consisted of atoms on a vastly smaller scale, which in their turn might be planetary systems, *ad indefinitum*. On this supposition, however, each electron would be a complicated physical object; it could be unlike all other electrons. Jupiter is heavier than the earth, Mercury is lighter, yet they are all planets; but electrons are much more alike than this. (Otherwise such things as television would not be possible. Electrons which varied in mass, even slightly, would never allow sharp images to form on the cathode screen. Indeed, on this principle one can ascertain that if electrons do differ in mass, they cannot do so by more than one part in 100,000. This is true also for atoms of gold, aluminium, etc.)

'But there is a difference between "do not differ in mass by more than one part in 100,000..." and "do not differ at all, are completely identical".' True. No empirical test can prove complete identity; more accurate observations may reveal minute differences. We can never say with certainty that two atoms or electrons are identical—not when our reasons for this lie only in experiment and observation. But this is equally true for classical physics: that two successive falls of the same sphere from the same height require *exactly* the same amount of time, could never be ascertained just from measurements.

'Then the question whether two particles are completely alike must be idle. If experiment and observation cannot decide the question, then it can never be decided.' Not at all. For the question, like others in this essay, is not an exclusively experimental one: it does not rest simply in the accumulation of more data. As before, this is largely a conceptual matter, one requiring reflexion, not the reiteration of tests. It has more to do with the pattern of quantum physical thinking than with particular experimental details which fall into that pattern. Naturally, that the details fall into the pattern profoundly affects our appreciation of both.

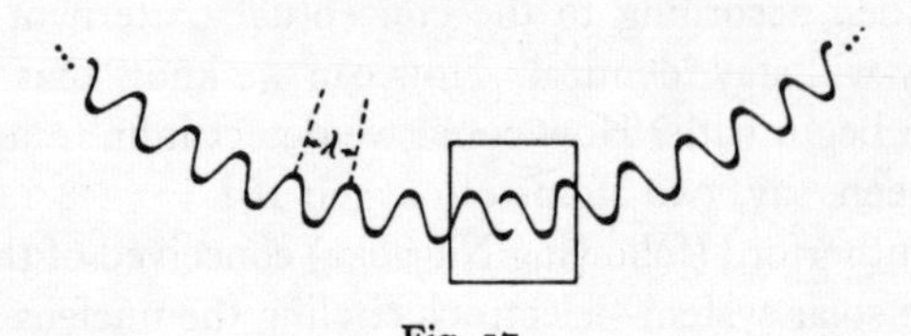

Fig. 17

Within quantum theory the treatment of any pair of interacting objects depends on whether or not they can be said to be completely alike. An oxygen molecule consisting of two equal and identical atoms behaves differently from a molecule of oxygen whose atoms are slightly different isotopes of oxygen, e.g. O_{16} and O_{17}. These differ in weight by a few per cent (O_{17} = isotopic mass 17·00450).

It was remarked in section B of this chapter that elementary particles are wave-like in certain respects. This will be examined, but for the moment let us concede that the waves associated with a particle's orbit join up smoothly, without discontinuity. On china plates on which a pattern repeats itself along the edge, it often happens that the pattern does not fit properly where it joins (see fig. 17); but this is never found in the wave behaviour which concerns the microphysicist. The wavelength, λ, must be equal to the circumference of the circle divided by some whole number N (i.e. the number of potential wave crests on the whole circle). If we tentatively visualize an electron thus:

at no stage in an electron's intra-atomic orbit would it be representable as in fig. 18; for here again,[1]

$$\lambda = \frac{2\pi r}{N},$$

λ can be a twentieth, or a nineteenth, or a twenty-first part of the circumference, but nothing in between. That it should be $\frac{2\pi r}{20.5}$ or $\frac{2\pi r}{21.1}$ is conceptually impossible: quantum thinking cannot allow this. Pianistic thinking cannot allow violinistic *glissandi*: pianos allow a C♯, or a D, but nothing in between. Classical physics regarded nature as a complicated violin: that is, differential equations were always in place; but we cannot think of the atom thus. λ depends on the speed of the particle, so a particle in an atomic orbit can only run at certain definite speeds. It cannot run at intermediate speeds; it cannot *exist* at intermediate speeds.

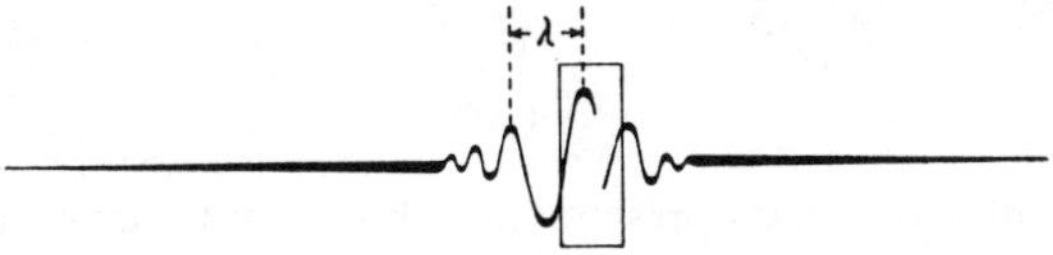

Fig. 18

Consider the molecule of hydrochloric acid, one atom of hydrogen and one atom of chlorine. The hydrogen atom is lighter, and swings around the heavier, near-stationary chlorine atom like a discus in the hand of the spinning athlete. It is as if the athlete could spin only at certain definite speeds: no faster, no slower, and nothing intermediate. Quantum theory was forced by observations to construe atomic dynamics in this way; it was driven to provide a mathematical technique for calculating these unique, determinate speeds. These calculations can be checked against laboratory measurements, such as those derived from spectroscopic study involving band spectra. The speed at which the hydrogen atom swings around the chlorine atom can be measured: all the speeds built into the pattern of quantum theory actually occur, but no others.

Suppose that in an oxygen molecule the atoms are equal in weight; they swing around each other, just 180° apart. Were they different in any way, each would have its own wave pattern with its own corresponding speed; but so long as they have the same wave pattern they are perfectly identical. When the speed is measured, however, it is found that not all the allowed values (odd and even) occur. Every other value, every odd one, is missing. That is, we detect 20 wave crests in an atomic orbit, or 22, or 24, or 26, but not 21, or 23, or 25, as seemed to be allowed in the early theory. Actually, this is precisely what must happen if the two oxygen atoms are completely alike, perfectly identical. It makes no difference if we turn the molecule through 180°, and the same must be true of the wave patterns concerned. A pattern with an even

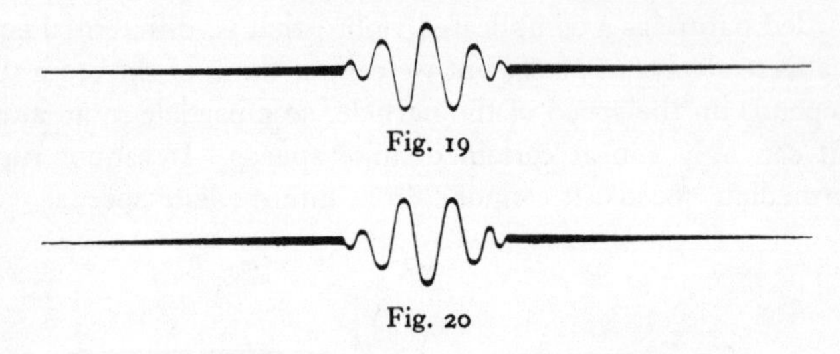

Fig. 19

Fig. 20

number of wave crests presents the same appearance at 180° as it does from 0°: it is like the lighthouse lantern which sends beams fore and aft, but sends the *same signal* out to sea twice for every one spin of the lantern. But if the number of wave crests in the orbit were odd—21, 23, 25—then at 180°, crests would be replaced by troughs; turning the pattern would change the appearance, would give a different signal. The wave pattern associated with one oxygen atom would be as in fig. 19: whilst that of its 'twin' would be as in fig. 20;[1] and the asymmetry would be immediately detectable. The slightest dissonance of this sort would be progressive, and (at the high r.p.m. of an oxygen molecule) would show clear discrepancies in band spectra. So the wave pattern of this oxygen molecule must have an even number of wave crests around the central point, and each atom must be identical with the other; otherwise wave mechanics collapses at its conceptual foundations. In an ordinary oxygen molecule, patterns with an odd number of waves cannot occur without transforming the molecule. This, of course, accounts for just the speed values which are found to be absent.

Laboratory experiment and measurements alone would never enable us to state the case with such finality. Thus it is not that the difference between two oxygen atoms is smaller than one part in 1,000,000,000,000: there is *no difference whatever*. If there is, then our present conceptions are totally and fundamentally wrong. But is the logic of this empiricist objection really clear? What possible alternative is being alluded to? If there *were* any difference whatever, a gradual rotation of the molecule could be detected; wave patterns with an odd number of waves would be possible and immediately obvious. Oxygen molecules in fact can be constructed in which the constituent atoms *are* slightly different: one is an ordinary oxygen atom, atomic weight 16, the other is the rare oxygen isotope (mass 17·00450) of which we spoke. Despite this small difference (6%) in measurement of velocity, all the allowed values occur, odd and even, i.e. 20, 21, 22, 23, 24, 25, 26, etc. with equal abundance. There is every theoretical reason for predicting that this would be so, however small the difference; nothing exposes discrepancies more quickly and more positively than superimposed wave disturbances.

There are alternative ways of determining whether two particles are identical, but the answer is always the same: they are completely identical unless they are obviously different. The guarantee for this is primarily conceptual. The enormous amount of experiment which supports the conviction is also made intelligible in terms of it. The experiments lend weight to the concept; but the concept was not precipitated *out of* the experiments.

'But why could there not be oxygen atoms which differ just slightly; where the effect of this difference remains below the lower threshold of sensitivity in our instruments?' Because these same arguments apply with more force to the particles[1] of which the atoms are composed. These too are identical. If these electrons, protons and neutrons, are conceptually standardized, then quantum theory puts it that they will always arrange themselves in exactly the same way. So the fundamental question is: Why are all electrons exactly identical, and similarly all protons, and all neutrons?

It is an indispensable condition of quantum theory that all electrons, all protons, all neutrons, must be identical; the successes of microphysics rest on this conception. If it is questioned, then all the achievements of two generations come under question as well. The theory allows us to calculate, from no more data than the number of electrons in an atom of some element, many of its physical and chemical properties. These predictions are supported in experiment. In principle, all of the properties of all the elements could be worked out; only fearful mathematical complexities prevent it. Before calculations can begin, however, certain properties must be 'worked into' the concept of the electron. Euclid had to accord certain properties to points and straight lines, before there was anything to calculate from; and similarly, the absolute identity of all electrons is a property they must have if they are to explain and pattern all the observations which gave rise to the electron-concept in the first place.

The power and success of quantum theory consists in the pattern of interlocked, systematic accounts it gives of the behaviour of complex bodies. Since it does this only by postulating the absolute identity of all elementary particles of the same type, what better reason could there be for saying that all elementary particles of the same type are identical? 'Then the identity of all electrons is just an assumption. It is true because physicists will not hear of it being false. It is a definition, pure and simple, hence arbitrary.' This will not do. Quantum theory is not mathematics: a conception like this would not have been formed *no matter what*. It is justified in every microphysical experiment, now and during many years past; indeed, without this conception experiments would not even make sense. All the data, the facts, the observations, bear the stamp of this unifying conception.

Why is a Euclidean point just the intersection of two one-dimensional lines? Why does a body free of impressed forces move in a straight line? Why does Mars describe an elliptical orbit? Why does the force of gravity vary inversely with the square of the distance? Why does nothing move faster than light? Why are all electrons identical? Because the world as we now know it becomes intelligible by supposing these things to be the case. What better reason for saying that they are the case?

New electrons are created when γ-rays pass through matter, and these are exactly like ordinary electrons. It may be objected: 'Insisting that all electrons are, for instance, the same size will not explain why this is so. Why cannot we imagine a *model* of an electron scaled up by an arbitrary factor? Unless we could imagine scale models of things we could not use the concept of size as something separate from the nature of an object. We can scale electrical voltage up or down in this way: one million volts is more difficult to manage than one hundred volts, but the physical laws are the same for both. We can even apply low voltage to a high voltage machine in order to study charge distribution. The answer will be independent of the voltage used.' 'Independent of the voltage used', but with an important proviso: the voltage gradient, that is, the rate at which the voltage varies from point to point, must not be *too* high, otherwise we would get a spark or a glow discharge. (It is as if Alice were to explode at a certain critical size.) This concept of electrical breakdown underlies all explanations of why electrons are, and must be, the same size. In the immediate locale of an electron the voltage gradient is a billion times greater than anything we can examine experimentally. Born explored the possibility that at such an enormous voltage gradient the laws of electrodynamics actually require modification (1934); his modified laws are designed to remain valid however closely the electron centre is approached. There is not really an electron here at all, not in the sense of being an infinitesimal, massy particle—it is more like a mathematical point imbedded in an incredibly strong electric field. This is the well-known *singularity*. Born demonstrates that only one kind of singularity is possible (at least in his theory); hence all singular points that do occur must be completely alike.

Anything which could not be described by something like Born's non-linear equations would just not be an electron; but any particle which *is* so describable must be identical with every other particle which could be so described. Observing electrons is clearly a sophisticated, theory-backed activity. It is observation of entities which are, as a matter of principle, unpicturable wave-particles, and absolutely identical.

We can now turn to a feature of the theory which has attracted philosophical attention, the uncertainty relations.

D

In einem bestimmten stationären 'Zustand' des Atoms sind die Phasen prinzipiell unbestimmt, was man als direkte Erläuterung der bekannten Gleichungen

$$Et - tE = h/2\pi i \quad \text{oder} \quad Jw - wJ = h/2\pi i$$

ansehen kann. (J = Wirkungsvariable, w = Winkelvariable.)[1]

Here is the first statement of the uncertainty principle. After thirty years it is still misunderstood, sometimes with philosophically disastrous consequences; in trying to allay these errors more of the conceptual framework of elementary particle theory may be revealed. In particular, the impossibility of at once precisely locating the position and measuring the velocity of an elementary particle will be shown to constitute not merely a technical impossibility, as many still imagine, but a conceptual impossibility. In elementary particle theory, phenomena can be appreciated only against a conceptual pattern one of whose features is an indeterminacy foreign to the thinking of classical particle physics.

This does not mean that the principle is an empty tautology, a definition, or an arbitrary notational convention. Had nature been other than it is, or had we come to conceive of it differently, the principle might never have been formulated at all. Our nineteenth-century professor did not himself distribute the spheres in the odd way we noted; nature was supposed to do that. What he did was to harness the only conceptions which could then describe what nature had done. Consider now three presentations of the uncertainty principle.

It might be said of the first of these, the Bohr-Heisenberg[2] super-microscope analogy, that it presents a technical impossibility, not a conceptual one. We may also say this of the second presentation, which arises out of actual experiments. But these two miss the mark: the first is merely illustrative, the second misleading. This is not always understood by those who appeal to them when describing the uncertainty relations. The third presentation is more challenging: it brings out the logic of the quantum-theoretic wave equation, and shows that the impossibility of absolutely precise observation is built into the concepts behind the wave equation—much as the possession of two kings is an impossibility involved in the concepts behind the game of chess.

The first presentation involves a *gedankenexperiment*; it supposes a super-microscope more powerful than any electronic microscope. We wish to determine the 'state' of an electron, its simultaneous position and velocity, so we 'magnify' the electron to visible size. However, the light which illuminates the electron must have a wave-length comparable to the dimensions of the electron itself, of the order of 10^{-8} cm., since light of a greater wave-length would not be reflected. Thus, our source of illumination must have the wave-length of X-rays (10^{-8} cm.) or shorter.[1] The shorter the illuminating wave-length, the more definite the image in the super-microscope; the sharper the image the more accurately can we locate the electron. We must, indeed, have a γ-ray microscope.[2] With this short wave-length of light, however, a disturbing phenomenon is encountered. When played upon an electron the X-ray or γ-ray behaves like a torrent of particles, and instead of illumination we get collision. Instead of our discerning the electron, it is bumped out of view altogether. This is the Compton effect, which constitutes part of the experimental basis for the view that light is granular or corpuscular, consisting in pulses, photons, free quanta. These photons are wave-like; photon beams diffract and interfere. But in collisions they behave purely as particles.

We had hoped to illuminate the electron. We have succeeded only in a classical two-particle system involving an elastic collision, as if we had bumped two billiard balls together. The initial velocity of our electron will have changed by an unpredictable amount. Discerning the position of the electron by this super-microscope has denied us any possibility of knowing its velocity accurately. In fact, given an electron of mass $= 9 \times 10^{-28}$ gm., whose position we decide to estimate with an accuracy of $1/100{,}000$ cm., our error in determining its velocity will be $(6 \times 10^{-27} \times 100{,}000)/9 \times 10^{-28}$, or $2/3 \times 1{,}000{,}000$ cm./sec., or 6 km./sec. If we reduce this uncertainty by using less energetic photons, that is, of longer wave-length, the image will become diffuse and the electron's position will be correspondingly less certain. This inability to determine the 'state' of an electron arises from having to illuminate it; the *gedankenexperiment* brings out what would happen were this attempted. Is it not reasonable, then, after *only* these considerations, to represent the accurate measurement of the 'state' of an electron as a technical impossibility, not as a conceptual impossibility?[3]

We may, however, find an alternative way of attacking the problem, one which involves no illumination in the sense described. We might find *some* way of tracing the particle through a weak field, in some manner which would not perturb the specimen. It is difficult to conceive what such an experiment would be like, for everything we know about elementary particles is learned by making them interact with familiar substances. It is this very interaction which perturbs the particle; a field so weak as not to perturb a test electron would not be an observable field at all. Still, it would be rash to regard present experimental techniques as final. Was not the wave nature of X-rays detected only after ordinary diffraction gratings were abandoned and a crystal used as a diffraction medium?[1] Did not this turn the technically impossible into the technically possible? Our hypothetical experiment involves techniques which allow no answer to the question 'What is the precise position and velocity of electron *e* at time *t*?'. Yet some way of providing an answer might be discovered; why bolt the door against this possibility?

There are several reasons why. The pattern of microphysical theory makes the possibility of such an experiment unfeasible. The super-microscope is hypothetical: there are no super-microscopes, nor could there be. Consider the modifications required in contemporary physics before the construction of such an instrument could become conceivable. Indeed, whether a 'γ-ray microscope' is even thinkable is too seldom recognized as raising issues of principle.[2]

Actual experiments are more like this: if we seek to determine the position of a particle precisely, we shoot a beam of electrons through a series of diaphragms, in each of which there is a narrow slit. The scatter of the emerging particles is then examined. In fact, crystals or crystal powders are used, and the 'slits' are their interatomic distances; but we shall talk of diaphragms. If we aim to determine velocity with exactness, the beam must pass through diaphragms in which the slits are wide, for the probability that a given electron will communicate part of its momentum to the diaphragm, by diffraction, is then very small. It cannot be said definitely of any electron on the 'far' side of a diaphragm whether it got there by passing through the slit, or by penetrating the matter of the diaphragm, or by bouncing off the inner walls of the slit.

Thus, whether the particle in fig. 21 came through the slit, or banked off its inner wall, or came through the diaphragm-wall, cannot be said with certainty; but naturally, if the last, the particle will have lost momentum in overcoming inter-atomic forces within the diaphragm-wall. We may eliminate this alternative by thickening the diaphragm, thereby lowering the possibility that any particle will actually get through; but this will still leave diffraction off the inner walls as a problem. We can say that the wider the slit the less likely it will be that the particle came to its position by reflection off the diaphragm-wall. If the electron did come through the slit, then its behaviour is essentially that of a Newtonian

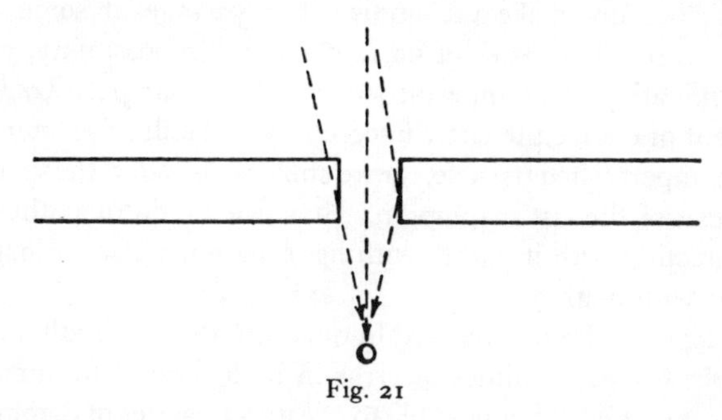

Fig. 21

'particle' moving in an inertial path; but then its position may be anywhere within an interval of possible positions—anywhere within the width of the slit. If the slit is narrowed, the position is more certain, the interval of possible positions is smaller, but the momentum possessed by the electron before reaching the diaphragm is less certain. For the probability has increased that the electron got to its final position by penetrating the diaphragm-wall or by diffracting on the slit's inner walls, instead of by-passing cleanly through the slit.

Which brings us to this: the precise determination of the state of an electron would involve the construction of a series of diaphragms, each one of which had a slit in it. But each slit must be both *wide and narrow* at the same time. Each diaphragm-wall, furthermore, must be at once *thick and thin*. It must be thick to make it less probable that any particle can penetrate it; this, however, makes the slit's inner walls longer, raising the probability of diffraction. Minimizing diffraction raises the probability of penetration. This is a conceptual *impasse*. What sort of technical advance could make possible the construction of such a diaphragm? There is no question here of techniques perturbing the specimen; nothing now corresponds to illumination bumping the particle out of position. What we have here are inverse relations between probabilities. As we try to close in on the electron's position, the probability that our narrowing slits will affect its velocity is raised. When we try to control this latter probability by thickening the diaphragm, diffraction becomes more probable. When we try to control *this* probability by widening the slits, the particle's position becomes less certain.

This does not make the undisturbed passage of some single electron through a series of narrow slits an impossibility, as was the illumination of an unperturbed electron in our *gedankenexperiment*; but of course the latter uncertainty will still arise later, even with an unperturbed particle, for we still have to *detect* the specimen at the exit of the last diaphragm. How can we do this other than by interacting with it, either with light, or some electro-magnetic field, or with matter?[1]

It is logically impossible within quantum theory simultaneously to reduce the probabilities of error in both these determinations, because it is logically impossible to construct a series of diaphragms which are at once thick and thin, and whose slits are at once wide and narrow. 'Is this different, however, from the super-microscope experiment? Why is this not a technical impossibility? We may hit upon a method of 'state' determination which does not require diaphragms. Why not?' The two *are* different. The *gedankenexperiment* is a physical account of why particles are perturbed when illuminated. Actual experiments, however, are summarized by saying that the degree of uncertainty in determining a particle's position is inversely related to the degree of uncertainty in determining (simultaneously) the particle's velocity. The 'mechanism responsible' for this need not be known, or even speculated about, for us to be able thus to describe observations. The diaphragm example need not be wedded to any theoretical account of the underlying physical processes. (Note also that 'a diaphragm slit cannot at once be wide and narrow' is necessary. 'The collision of a photon with an electron will perturb the electron' is not necessary.)

Still, there is no more finality about diaphragm experiments than about any other technique. Our two presentations are further distinguished by the assumption that in both, the electron is a point-particle which 'really' has a unique state; only our observing techniques make it impossible to assess this. Whatever the new techniques, however, they will not allow us precisely to determine the simultaneous position and velocity of an electron. They cannot —not within the conceptual framework of quantum theory. The essential consideration has not to do with thinking about techniques of measuring matter, but arises from thinking about matter itself. Unless the whole of quantum theory is discarded, uncertainty is here to stay;[1] it is built into the conceptual pattern of quantum mechanics. Uncertainty is not something discovered by experiment in the sense that one 'winds up' the apparatus and can then observe the uncertainty relations. It is nothing encountered as an experiment-datum, yet every observation in microphysics is what it is because of these relations. The uncertainty principle is not a detail of microphysics, it is an essential part of the plot. It patterns microphysical phenomena for the physicist; it is not just an awkward anomaly, as some suppose. The pattern was built up by studying such phenomena, but it is not itself one of those phenomena.[2]

After rolling his test spheres as shown in fig. 14 (p. 127) our nineteenth-century professor gets the totally unexpected result of fig. 16. After many efforts to 'correct' his readings, he must concede that such a distribution could only result from interference; besides their particle properties the spheres show wave behaviour. To describe this adequately, one must combine in one physical language two notations, a particle notation and a wave notation. What would this be like?

The classical concept of a particle is of a dimensionless point endowed with mass—Democritus' atom. Any particle P at any time has an exact spatial position and velocity; P is the 'massy intersection' of four co-ordinates.[3] Thus, from the statement that P is a particle we may conclude that *P has a precise and determinable spatial position and velocity*. Let us call this 'Q'. The classical conception of a wave, however, entails that no wave W can have a precise spatial position at a specified time.[4] Ideally, a wave disturbance spreads itself all through the medium in which it occurs; but at a geometrical point a wave disturbance is no disturbance at all, simply a periodic pulse. With the 'wave packet' the situation is similar: the 'spread' of such a packet is an essential part of its having recognizable wave properties. Therefore, from the statement that W is a wave (or a 'packet' of waves) we may conclude that *it is not the case that W has a precisely determinable spatial position.* (Where W is a wave packet it has no uniquely determinable velocity either.) Let us call this '$\sim Q$'.

So far $\sim Q$ above is not the negation of Q; Q refers to P, while $\sim Q$ refers to W. However, when 'P' and 'W' are used jointly to designate the same entity, the point of the '$Q. \sim Q$' notation will be apparent. Our professor must represent any given sphere as a PW, since it possesses both particle and wave properties: that is the conceptual situation his observations have forced on him. Let us represent 'PW' by 'Ψ'. We must now say of any one of our hypothetical spheres that (1) it has a precise and determinable spatial position and velocity, and (2) it has *not* a precise and determinable spatial position. In other words any Ψ is such that $Q. \sim Q$.[1] This is a logically intolerable situation for our nineteenth-century professor. Initially, the experiment will be repeated many times—long after his undergraduates leave. Signs of mismanagement or miscalculation will be sought; he will explore every possibility of saving the concepts of his physics, hunting (for instance, in the boundary conditions) for systematic errors or undetected forces. Everything else undergoes scrutiny before the concepts which pattern an experiment are examined.

The results are always the same: it gradually dawns on the professor that this is no accident. 'Is it necessary, however, that the two notations be combined in this way? Some other form of expression might embrace these findings without the logical embarrassment of assigning properties which are incompatible.' Our classical conception of *particle* and *wave* being what they are, *some* sort of combined notation is inevitable. What other alternative is there? By what other route is one likely to find an *explicans* for the astounding distribution in fig. 16? Is $Q. \sim Q$ merely to be shrugged off? Or would increased familiarity with this painful consequence harden him to it?[2] Surely not; for conceptual equilibrium outlaws $Q. \sim Q$. Our professor would probably do as did his followers generations later: he would combine the only two languages available, but with the restriction that '$Q. \sim Q$'

cannot be formed; questions which invite this as an answer are ruled out.[1]

The professor would continue to speak in an approximate way of the position and velocity of his Ψ spheres, but, because of his discovery, he would soon realize that talk about Ψ position and velocity could not be precise beyond a certain limit. Notice how little of this conclusion depends on any low-level experimental facts. Granted, the whole business is 'triggered' by the observed sphere-distribution; but a purely formal exercise requiring combination of wave and particle notations would give the same result. From the mere manipulation of these notations one could draw out something like the incompatibility we have noted. Our classical conception of a wave entails that it 'spreads' in space and time. This limits the accuracy with which we can describe anything characterized as Ψ, for an absolutely precise point-location would eliminate the W phase of the notation; it would eliminate 'spread'. Conversely, an absolutely precise estimation of momentum would eliminate the P phase; the 'spread' would have become infinitely great. Any mathematician after Fourier could have shown this without special empirical knowledge. To insist on Ψ is to make absolute precision for both P and W impossible. This suggests a sense in which the uncertainty relations may be said to be built into our *classical* concepts of particle and wave.

What does rest on empirical data, on physical experience and insight, is the initial decision to combine concepts in this manner. Nothing less than a discovery as unsettling as the one depicted could force so difficult a decision. This is of a type with Kepler's decision about a non-circular Martian orbit, Galileo's introduction of t into discussions of falling bodies, and Newton's resolution to ignore the causes of gravity.

Our nineteenth-century professor would regard as ill-conceived the question 'Yes, but what *is* the precise state of the sphere Ψ at time t; you cannot deny that it *has* such a state?'. To answer this as it demands would force the conjunction of Q and $\sim Q$, or the abandoning of one phase of the Ψ notation. Neither alternative is feasible. The world being what it is (i.e. the spheres-distribution being as shown), and our conceptions of particle and wave being what they are, there is no point in talking of any Ψ as having a precise position and velocity. The professor's concepts could not link up with such a question; he can no longer see the spheres as he had done before observing this distribution. Nor can he now state the 'facts' of the pre-Ψ physics of rolling and scattering balls.

'Well, this professor never existed in the nineteenth or any other century. No such experiment was ever performed. Spheres would not actually distribute themselves as above. No nineteenth-century physicist ever had any such problem; nor was any such restriction ever made.' True; but this *gedankenexperiment* brings out conceptual features of the very situation in which physicists of this century did find themselves. Quantum physicists agree that subatomic entities are a mixture of wave properties (W), particle properties (P), and quantum properties (h). High-speed electrons, when shot through a nickel crystal or a metallic film (as fast cathode-rays or even β-rays), diffract like X-rays. In principle, the β-ray is just like the sunlight used in a double-slit or bi-prism experiment. Diffraction is a criterion of wave-like behaviour in substances; all classical wave theory rests on this. Besides this behaviour, however, electrons have long been thought of as electrically charged particles. A transverse magnetic field will deflect an electron beam and its diffraction pattern. Only particles behave in this manner; all classical electromagnetic theory depends upon this. To explain all the evidence electrons must be both particulate and undulatory. An electron is a PWh. This is Ψ, with Planck's quantum of action worked into it. We still call it 'Ψ'.[1]

Consider Louis de Broglie's formula $\beta = h/mv$.[2] 'λ' symbolizes wave-length, as in classical wave theory, 'mv' symbolizes momentum in the Galilean-Newtonian 'particle' sense and 'h' is the quantum constant of proportionality. Here in one terse expression are all the notions required for generating elementary particle theory.

If v is the particle velocity, as indicated in experiments, what is the wave velocity? Since the wave motion and the particle are intimately associated, the wave must 'move along' with the particle, and v must be its velocity also. 'But waves do not just transport themselves from place to place as physical objects. Waves are continuous disturbances in a homogeneous medium, periodic in space and time. A continuous disturbance filling space is hardly a suitable particle model. It is not localized as particles must be. How then is a wave to be localized?'[3] Fig. 22 shows two waves.[4] At A these waves are out of phase, so the resultant wave motion at that point would be nil; each would cancel the effect of the other. At *B* the two waves are in phase, so resultant wave motion there would have double the amplitude of either component wave. The waves interfere destructively at *A*, and constructively at *B*. If we add other waves of different wave-lengths which do the same, we get the result shown in fig. 23.[1] This is an interference pattern, fixed diagrammatically in space. It resembles a whistle blast or a light flash. We characterize a whistle blast by a set of sound waves of various wave-lengths ranging around some one average value; if all these individual waves are mapped and cancelled and augmented according to the laws of constructive and destructive interference, the resultant pattern is as in fig. 23.

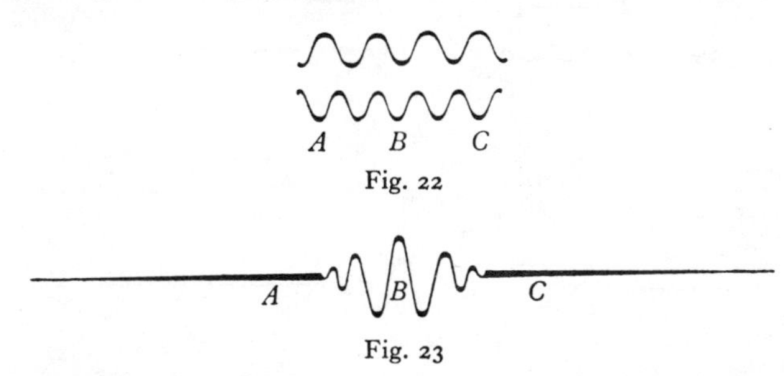

Fig. 22

Fig. 23

With an elementary wave-particle, however, the pattern is not a split-second blast or flash: in an important sense the electron endures, and its endurance and its motion are intimately connected. The pattern progresses in the direction of wave propagation as a pulse, or a packet of these interference maxima: at any instant these maxima are representable as a fixed interference pattern. The presence of an infinite range of wave-lengths, however, makes it as if the maxima moved in the direction of the waves. It is as though the disturbance appeared on each successive square of a cinema film a little advanced to the right. In fig. 24, wave maxima 1, 2 and 3 are like the one earlier. In a way, wave pulse 1 has ceased to exist before 2 is formed. And as wave pulse 2 'collapses' all its constituent waves re-configure so that maximum 3 comes about. The resultant motion is representable as the classical motion of a single pulse, as Schrödinger[2] and many others have shown (fig. 25).

The movement of a wave-packet is thus like an enormous number of dolphins crowding just below a still surface. Their jostling pushes one dolphin above the surface, then another and another, so rapidly and so closely that it might be a single dolphin flashing across the water, according to the laws of classical particle physics.[1] Similarly the wave maxima 1, 2 and 3 are 'pushed' into being by a swarm of continuous wave disturbances, and each maximum is just a little farther along in the direction in which the whole school moves. This is a crude picture; but so, when it comes to that, are all the wave-diagrams. These illustrations are ladders we shall certainly throw away once they are climbed.

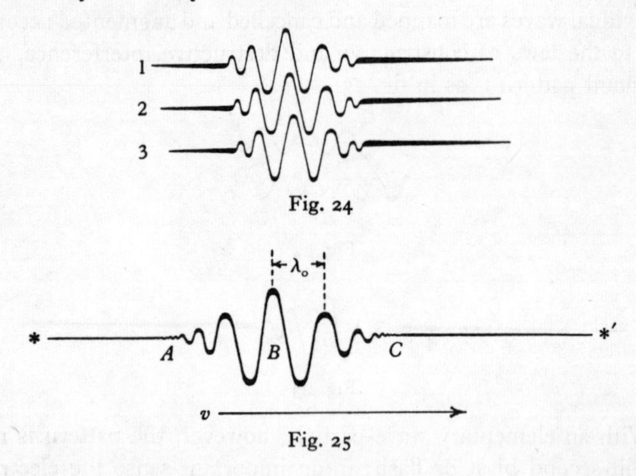

Fig. 25

So 'v' is the speed with which each newly created pulse Ψ appears, in the direction of wave-packet propagation. Each Ψ is identical with each other Ψ. This is not like a series of very similar dolphins surfacing one after another; it is like absolutely identical, indistinguishable dolphins following one another. Were there any difference whatever between Ψ_1 and Ψ_2 (other than in co-ordinates), they would not both be represented by 'Ψ'. In the last diagram, λ is the average wave-length in the packet. Outside the packet, at * and *′, there is complete destructive interference of the component waves. (The ocean's surface is unrippled here as the dolphins nullify each other's efforts to 'break water'.) This results in a localization of the wave-packet. At the centre of the group all waves are in phase and give a maximum effect; here the voltage gradient, mentioned earlier, is fantastically great. This permits description of an elementary particle as such a wave packet, Ψ.

The particle velocity is the wave-packet (Ψ), velocity v; the De Broglie wave-length λ is some one of the wave-lengths in Ψ. We cannot say which wave-length, however; and that is the fundamental complication. In the propagation of light, waves of various wave-lengths move with the same velocity (disregarding dispersion, that is); this is the case with sound waves too. Thus the pulse velocity and the ordinary wave velocity are the same. A shaft of white light hits us all at once, not as a spectrum. But in the De Broglie formula, λ is dependent both on the pulse velocity v and on the ordinary wave velocity. There is a spread in λ from A to B and

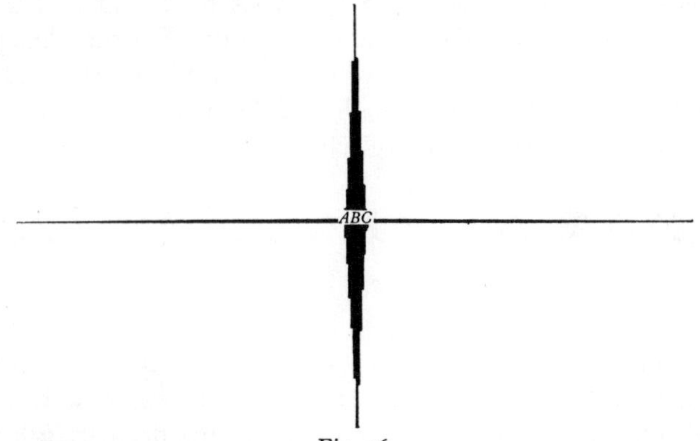

Fig. 26

from B to C, and there is a spread in the ordinary wave velocities in the pulse. No exact correlation between particular wave-lengths in Ψ is thus possible.

So the wave-packet model of an atomic particlé is inherently 'fuzzy'. Schrödinger originally put it that the electron is not a localized point charge at all—a step De Broglie did not actually take—its charge and mass are *smeared* over a certain region. (Though Schrödinger was wrong about this, there is an advantage in adopting his exposition provisionally.) For a given wave-length there is no particular particle velocity, since this depends on the whole set of wave-lengths composing the pulse; and there would be no pulse unless all the component frequencies were different. Suppose we wish to determine precisely the position of an electron Ψ. We must eliminate the lateral 'spread' or 'fuzziness' of the particle; we must make the packet shorter. Making Ψ shorter from

A to *C* means adding more high-frequency waves—that is, intensifying the field of which the electron is the quantum aspect. Fig. 26 shows what we are after. Each of the new component waves must have a different wave-length if we are to have a high frequency, high amplitude packet. It follows that we will have increased the number of differing component wave-lengths between *A* and *B* and *B* and *C*. The mean difference between λA and λB of the diagram above will be much greater than that between λA and λB in the earlier one; there will be a spread of velocities as well. According to the De Broglie formula, each wave-length has a different wave velocity. The more localized the wave-packet—the closer *A* and *C* are to *B*—the greater this spread of velocities. For complete precision in position the velocities will spread over all possible values; complete knowledge of the position of the particle thus destroys all possible knowledge of its velocity.

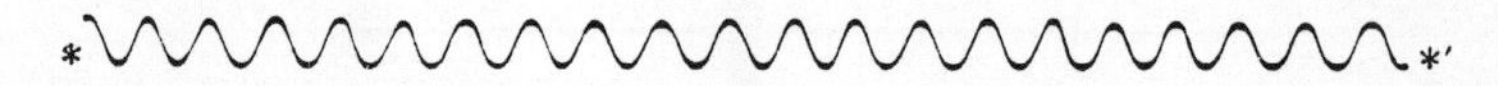

Fig. 27

Try the reverse: eliminate waves from Ψ until the wave velocity is known with precision. Fig. 27 shows what we are after now; this is one of the component waves of our first diagram, *une onde de phase*. Its velocity can be assessed very accurately, but eliminating waves minimizes destructive interference at $*$ and $*'$. It lowers the intensity of the field at *B* so that the 'particle' is no longer localized. The packet Ψ has now spread out: ultimately it will fill all of space. Complete knowledge of the velocity of the particle, then, destroys all knowledge of its position.[1]

Alternative presentations of this situation, for example that of matrix mechanics, issue in exactly equivalent results.[2] But an exposition of the conceptual pattern of matrix mechanics would be a formidable business, so we have chosen to broach the subject via wave mechanics, as is usual, even though virtually all that we shall retain are the two equations of De Broglie and Schrödinger—and even then only in a formal way. It does not matter what instruments we use: they may be the crude diaphragm-series, metallic foils and crystal powders of today, or the precise tools of the next century. It is a logical feature of this conception[3] of an elementary particle Ψ, that the simultaneous measurement of its position and velocity cannot be carried out with precision. The wave equation and its consequence, the uncertainty principle, may of course be given up,[1] but this would not be a reshuffling of one or two elements at the top of the pile of microphysical knowledge: the whole structure of that pile would collapse. It would not be a modified detail, but a thoroughly rewritten plot, an overhauled conceptual pattern: the better part of quantum theory would have been given up.

There are plenty of technical obstacles for the quantum physicist to hurdle: most of these he attacks from within the conceptual framework of the theory. The uncertainty principle is no such obstacle, for it is built into the outlook of the quantum physicist, into every observation of every fruitful experiment since 1925. The facts recorded in the last thirty years of physics are unintelligible except against this conceptual backdrop. One cannot maintain a quantum-theoretic position and still aspire for the day when the difficulties of the uncertainty relations will have been overcome. This would be like playing chess and yet hoping for the day when the difficulties of possessing but one king will have been overcome. To hold a quantum-theoretic position just *is* to accept the relations as unavoidable.

The whole theory may topple; in places the foundations seem far from secure.[2] Nonetheless we cannot see the micro-world as we now do without accepting the uncertainty relations as inextricable in the organization of what we encounter.

E

These reflexions become intense when we consider the correspondence principle which, on one interpretation, must be continually at tension with the uncertainty principle. Weyl says: 'Thus we see a new quantum physics emerge of which the old classical laws are a limiting case, in the same sense as Einstein's relativistic mechanic passes into Newton's mechanic when C, the velocity of light, tends to ∞.'[3] This is now a familiar kind of pronouncement Treatises in theoretical physics intend something special when they describe the correspondence principle in this way, and in such contexts one is rarely misled.[4] But in other contexts misconceptions can arise. Weyl's words could, for example, lead one into the following perplexity.

(*a*) The motions of planets, Mars for example, are described and explained in terms of 'the old classical laws'. These descriptions proceed as follows: while in practice it is not possible to determine the state of a planet by absolutely 'sharp' co-ordinates and momentum vectors, still, it is always legitimate to speak of it as having exact co-ordinates and momenta; this is always an intelligible assertion. In classical mechanics uncertainties in state determination are in principle eradicable. Expositions refer to punctiform masses, the paradigm examples of mechanical behaviour. Point-particles are distinct possibilities within classical particle physics.

(*b*) Elementary particle physics presents a different logical situation. As we saw, the discoveries of 1900–30, if they were to be explained, forced physicists to combine concepts in unprecedented ways, e.g. $\lambda = h/mv$. A direct consequence of these combinations of concepts is expressed in $\Delta p . \Delta v \cong h/m$, where Δp and Δv are the limits of uncertainty in a particle's co-determined position and velocity. Within quantum theory, to speak of the exact co-ordinates and momentum of an elementary particle at time t is to make no intelligible assertion at all. What could the assertion consist in? That a wave packet has been compressed not to a line but to a point? This cannot even be false, since one must at least have a clear concept of x to be able to use it in making false statement. Is there any clear concept of a wave-packet at a point? No. Again this is not simply the discovery that our instruments are too blunt for the delicate task of observing the simultaneous positions and momenta of microparticles. In the well-established language of quantum theory a description of the exact 'state' of a fundamental particle cannot even be formulated, much less used in experiment. It is, for instance, a condition of Dirac's theory that position and momentum operators are non-commutative; to let them commute is not to express anything in Dirac's theory.[1] Whatever the wave equation ($\Delta\psi + 8\pi^2 m/h^2(E-U)\psi = 0$) can be said to express, it cannot be 'squeezed' to a geometrical point; not without phase velocities spreading over all possible values. Nor can momentum be specified by a unique number without the positional co-ordinates being 'smeared' through all space. So if the Schrödinger equation is fundamental to the language of quantum theory, then, for anything which could be described by the ψ function, nothing like $v = ds/dt = \dot{s}$ or $a = dv/dt = \dot{v} = d^2s/dt^2 = \ddot{s}$ can obtain.[1] Point-particles, therefore, are not possibilities within elementary particle physics. However,

(*c*) Quantum theory embraces classical particle physics. 'We see a new quantum physics emerge, of which the old classical laws are a limiting case....'[2] The justification for this is as follows. The orbital frequency of the electron in a hydrogen atom is given by $\omega/2\pi = \gamma_{(cl)} = 4\pi^2 me^4/h^3n^3$. According to the classical connexion between radiation and electrical oscillation, this is the same as the radiated frequency. But quantum theory gives

$$\gamma_{(qu)} = (2\pi^2 e^4 m/h^3) \times (n_i^2 - n_f^2/n_i^2 n_f^2)$$

for the frequency of radiation connected with the transition $n_i \to n_f$. If $n_i \to n_f$ is small compared with n_i, we can write instead

$$\gamma_{(qu)} = (4\pi^2 e^4 m/h^3 n_i^3) \times n_i - n_f.$$

Thus, in the limiting case of large quantum numbers, $\Delta n = 1$ gives a frequency identical with the classical frequency, i.e. $\gamma_{(qu)} = \gamma_{(cl)}$. The transition $\Delta n = 2$ gives the first harmonic $2\gamma_{(cl)}$... and so on.

(*d*) Here the perplexity arises. A certain cluster of symbols S expresses an intelligible assertion in classical mechanics, yet that same S may not be so regarded in quantum mechanics. Could d^2s/dt^2 be translated into Dirac's notation? or von Neumann's? No; nonetheless the languages of the two systems are said to be continuous. Just as the law of inertia is said to be only a special case of the second law of motion, so classical mechanics as a whole is said to be but a special case of quantum mechanics. These are apparently just clusters of statements in the same language.

Statements and languages do not work in this way, however. A well-formed sentence S, if it can make an intelligible empirical assertion in one part of a language, must be capable of doing so in all parts of the language. Technical notations are usually *defined* in terms of rules which determine those combinations of symbols which can be used to make intelligible assertions. If S can be used to express an intelligible statement in one context, but not in another, it would be natural to conclude that the languages involved in these different contexts were different and discontinuous. Finite and transfinite arithmetics, Euclidean and non-Euclidean geometries, the language of time and the language of space, the language of mind and the language of brain; these show themselves to be different and discontinuous on just this principle. What can be said meaningfully in one case may express nothing intelligible in the other. This also happens when S expresses the state of a particle in classical physics—where the result is an intelligible assertion, as against S expressing the state of a particle in quantum physics (e.g. in Dirac's notation)—where the result is no assertion in that language at all.[1] Ordinarily this would be conclusive evidence that the languages are different, *logically discontinuous*. But the correspondence principle apparently instructs us to regard them as continuous; quantum theory embraces the old classical laws as a limiting case.[2]

This is a genuine perplexity. How can intelligible empirical assertions become unintelligible just because quantum numbers get smaller? Conversely, how can unintelligible clusters of symbols become meaningful just because quantum numbers get larger? The intelligibility of assertions within a language cannot be managed in this way.[3] A spectrum of intelligible assertability through which a single formula S can range within a language is unthinkable. Either S can make an intelligible empirical assertion in all of the language in which it figures, or else the latter is really more than one language. Either the uncertainty principle holds: that is, the S of classical physics makes no assertion in quantum physics, or the correspondence principle holds: that is, the S of classical physics is a limiting case of quantum physics; but not both. Or else we are misinterpreting one, or both, of the principles in question. First we are warned that the new physics is logically different from the old, and that we should not make old-fashioned demands on it. Then we are told that the two are quite harmonious. This clearly needs sorting out.[4]

The difficulty can be expressed in terms of probability distributions. Classical theory allows joint probabilities of accuracy (in determining pairs like time-energy and position-momentum), and allows them to increase simultaneously. In quantum theory this is illegitimate. But as quantum numbers get larger, the legitimacy of these joint probabilities seems to increase; the same perplexity arises.[5]

(*e*) So the alternatives seem to be: (1) quantum physics cannot embrace classical physics as a limiting case, or (2) quantum physics ought not to be regarded as permanently restricted such that no analogue for the classical S is constructable, or (3) classical physics itself should be regarded as restricted against the construction of S, just as is quantum physics.

Alternative (3) may be dismissed. It may, as Boltzmann remarked, be more faithful to the limitations of actual observation and experiment, but it is a self-denying ordinance of little practical value. A classical mechanics without punctiform masses would be conceptually too difficult to justify the recommended change.

Alternative (2) has been chosen by many eminent physicists and mathematicians: Einstein, Podolsky, Rosen, De Broglie, Bohm, Moyal, Bartlett and Sir Harold Jeffreys, to name a few. Thus: 'The limitations expressed by the leaders of quantum theory are not essential to the theory and arise simply because the theory has not yet been expressed in a sufficiently general form.'[1] This results from noting the tension between the correspondence principle and the uncertainty principle. Considering further the orbital frequency of the electron in the hydrogen atom, plus thirty years of uneasiness caused by the uncertainty principle, alternative (2) begins to look plausible. Nonetheless it rests on a misconception as to the nature of the correspondence principle.

The uncertainty relations are an intrinsic feature of quantum theory. As we saw, they are built into

$$\lambda = h/mv \quad \text{and} \quad \Delta\psi + 8\pi^2 M/h^2(E-U)\psi = 0.$$

These relations were implicit in the first decisions of De Broglie and Schrödinger to weld particle and wave notions into a single notation. Nor is Jeffreys' contention clear: how exactly could any mathematical generalization change the relationship between these two logically discontinuous ideas?[2] There is no ultimate logical connexion between the languages of classical physics and quantum physics, any more than between a sense-datum language and a material object language. This cannot be supported by appealing now to $\Delta p . \Delta v \simeq h/m$; that would be a *petitio principii*.

But let us remember at this point that classical particle physics is a dynamics of bodies set in a 'Euclidean' space-time framework. The order of development in the subject is always from kinematics to dynamics. We study first Galilean reference frames, vector analysis, and the properties of bodies at rest; only after that are dynamical ideas introduced into this geometrical framework. Naturally, points are just the intersections of one-dimensional co-ordinates. Punctiform masses are Euclidean points, in time, endowed with mass. Given two of these point-masses, most of classical dynamics can be worked out. With three of them some of the most complex problems in physics already arise.

If anything, development in quantum theory reverses this. Actually no such division is even possible; if it were, an elementary particle's 'kinematical' properties would depend on its dynamical properties and not vice versa. The Nagaoka, Rutherford and Bohr conceptions broke down just here: they tried to work new dynamical properties into the traditional framework. The subsequent difficulties are well known. De Broglie noted that to give a velocity *c* to a particle of mass $> O$ would require an infinite amount of energy. He asked whether such particles might be related to a wave mechanism as photons are related to the wave nature of light. Here is the starting-point of a new pattern of concepts: a wave motion at a geometrical point is inconceivable, and hence photons and electrons must 'spread'.[1]

The punctiform mass, a primarily kinematical conception, is the starting point of classical particle theory. The wave pulse, a primarily dynamical conception, is the starting-point of quantum theory. Languages springing from such different stock are likely to show this logical difference throughout; and this is indeed the case.

'What about the hydrogen atom with large quantum numbers? What is the explanation of that?' This has been misunderstood too. Languages of so different a conceptual structure cannot *simply* mesh in this way; their logical gears are not of the same type. Identically structured sentences and formulae, though they can express many different types of statement (cf. ch. v) cannot express single statements whose sense and intelligibility varies simply with the size of quantum numbers—not unless the statements are really set in different languages according to different rules; i.e. are different statements.[2] Propositions get their force from the whole language system in which they figure. That $(4\pi^2e^4m/h^3n_i^3)\times(n_i-n_f)$ gives a 'classical' frequency for the transition $\Delta n=1$ proves at most that there are formal analogies between certain reaches of quantum theory and certain reaches of classical theory. That there is simply an analogy can be obscured by the fact that the same symbols, '$4\pi^2 e^4 m / h^3 n_i^3$', are used in both languages; but this no more proves a logical identity between the two than the uses of '+' and '−' for both valence theory and number theory shows these theories to have an identical logic. The same point was noted in the last chapter in discussing the laws of Newtonian mechanics.

Men are made of cells. I might urge that whereas it is true to assert that men have brains, personalities, financial worries, it is no assertion at all to say these things of cells. This would be incorrect, for it would surely be an intelligible assertion, only false. Suppose, however, that cell-talk were made logically different from man-talk. The two idioms could never merge. 'It has schizophrenia and an over-draft' would not express an intelligible statement in cell-language. Even though a certain complex combination of cells could be spoken of in ways analogous to the ways in which we speak of men, this would not run the two languages together; not even when both idioms were directed to characterizing the same object, say myself. If you speak of me as a man, but someone else speaks of me as a collection of cells, then, though the *denotatum* of your talk be physically identical, the two of you diverge conceptually. You are not speaking the same language.

Similarly, in an intricate sense-datum language it might be possible to construct sentences analogous to material object sentences. In the same conditions, were it true to make a certain sense-datum statement it would also be true to assert its analogous material object statement. If when it is true to say 'I am aware of a brownish, grizzloid, ursoid patch', it is also true to say 'There is a bear before me', then the two sentences are analogous. This does not prove the *identity* of the two language systems. 'I see a bear before me' could be intelligibly asserted to express a false statement even when it is stated sincerely. Could one intelligibly assert this of 'I am aware of a brownish, grizzloid, ursoid patch' when it is sincerely stated? There are points of contact with respect to the application of the two languages, but this no more 'reduces' one to the other than does the language of mind (memory, sensation, character, habits, etc.) simply reduce to the language of brain. Nor does the language of Picasso reduce to that of Einstein when they are both speaking truly, in their special ways, of a sunset. Their utterances may be identical ('It is red now'), but their assertions may diverge widely. Similarly, 'The probability that *a* is a *b* is 1' is analogous but not equivalent to 'All *a*'s are *b*'s'.[1] The differences in these language systems are not changed by the fact that such utterances can often be made truly in the same context.

The logical continuity, suggested by careless statements of the correspondence principle and by examples concerning the energy levels of the hydrogen atom, is illusory. It does not show classical particle physics to be a limiting case of elementary particle physics, even though the formalisms of these two systems may be completely analogous at points. What it does show is that when quantum numbers are high the hydrogen atom can justifiably be regarded from either of two points of view; as a small macrophysical body set in classical space-time, wherein *S* will be an intelligible assertion, or as a large 'quantum' body exemplifying only to a small degree the dynamics of elementary particles, where *S* will not constitute an intelligible assertion.[2]

We are to some extent free to treat the hydrogen atom as we please, depending on our problem. Similarly, we treat Mars sometimes as a punctiform mass, sometimes as a solid oblate spheroid. We regard gases sometimes as dense, continuous media—in acoustics, for example, and sometimes as porous, discontinuous swirls of particles—in statistical thermodynamics, for example. The conceptual differences here reflect earlier examples: Kepler and Tycho at dawn, Beeckman's problem and Descartes' problem, Kepler's ellipse as a mathematical device and (later) as a physical theory, Kepler's laws as they were for him and then for Newton, as they came to be for Mach and then for Hertz. The hydrogen atom *qua* small macroparticle is as different conceptually from the hydrogen atom *qua* large microparticle as any of the differences in these examples.[3]

If one insists on a crude statement of the correspondence principle then the modification must be made in classical, not in quantum mechanics: the electron as a point-particle in Euclidean space is no *explicans* for the phenomena encountered in this century. One could, however, restrict celestial mechanics so that $\Delta p . \Delta v \cong h/m$. (Observationally, it was never entitled to punctiform masses anyhow.) This restriction would make no practical difference, for just as utterances concerning temperatures less than -273° C. now make no intelligible assertion, so then utterances concerning any body's exact state would be regarded as making no intelligible assertion. This is the recommendation that the derivative be regarded as lacking physical application. It is like the ruling that physical thinking should concern itself only with observable quantities; this is the recommendation of Einstein that we abandon talk about the aether and simultaneous interstellar events, and the recommendation of Heisenberg (as against Schrödinger) that elementary particle theory should abandon talk about individual electronic orbits, frequencies, velocities, etc., restricting itself to the scattering properties (matrices) of cathode, β- and γ-rays, and other 'group' phenomena. But all this seems unnecessarily Procrustean; classical mechanics is simpler as it is now.

There is no logical staircase running from the physics of 10^{-28} cm. to the physics of 10^{28} light-years. There is at least one sharp break: that is why one can make intelligible assertions about the exact co-ordinates and momentum of Mars, but not about the elementary particles of which Mars is constituted. As an indication of how the mathematics of elementary particle physics can be managed, the correspondence principle is clear and useful.[1] But when spoken of in more spectacular ways, as, for example, by Weyl, the nature of intelligibility in physics hangs in the balance.

F

Our concern has been not with *giving* physical explanations, but with *finding* them. How one employs accepted theories to explain familiar phenomena falling under them has not been the issue; rather we have tried to explore the geography of some dimly-lit passages along which physicists have moved from surprising, anomalous data to a theory which might explain those data. We have discussed obstacles which litter these passages. They are rarely of a direct observational or experimental variety, but always reveal conceptual elements. The observations and the experiments are infused with the concepts; they are loaded with the theories. When the natural philosopher is faced with the types of problem which we have been describing, his observations and his experiments will contain that problem. Galileo's geometrical bias,

Kepler's perfect circle, Priestley's phlogiston, Leverrier's unobserved masses—these were the obstacles, so say we now with the wisdom of hindsight. We cannot so confidently judge the ideas which control physical thinking today, such as about uncertainty and the Ψ function, but we *can* note again that these 'obstacles' affect the entire organization of one's data, observations, facts and subsequent theories. Nor is this true only of those who are stopped by the obstacles. Only by seeing what sorts of things make a man fail to explain a phenomenon or fail to make a certain observation can we appreciate what is at work when he succeeds at these things.

Intelligibility is the goal of physics, the fulfilment of natural philosophy; for natural philosophy is philosophy of matter, a continual conceptual struggle to fit each new observation of phenomena into a pattern of explanation. Often the pattern precedes the recognition of the phenomenon, as Dirac's theory of 1928 preceded the discovery of the positron, the anti-proton and the anti-neutron, and as Pauli's neutrino-hypothesis preceded the actual discovery of the particle by more than a generation. But then Dirac's pattern was itself the issue of an effort to find a suitable *explicans* for prior phenomena, i.e. a unified, relativistically invariant theory of electron spin which would give the correct fine structure formula, the Zeeman effect of the doublet atoms, a description of Compton scattering and a model of the H-atom. This resembles the way in which the anomalous behaviour of Mercury at perihelion vexed all Newtonians after Leverrier, being made intelligible only in 1915 by general relativity. It is like Newton himself struggling up from Kepler's three laws—apparently distinct and independent—to universal gravitation, which gave these laws a coherence and integrity they never had for Kepler. The planetary-ellipse in its turn had done the same thing for Kepler's thinking about the chaotic heap of Martian data he had inherited from Tycho Brahe; Kepler's observations of Mars acquired something that those of Brahe and Galileo never had. And for the Galileo of 1638 freely falling bodies were not what they had been for the Galileo of 1604. These were all great advances in physical science. In principle they are like the advances Tycho and Simplicius would have made had they come to appreciate the sunset as did Kepler and Galileo. 'Though the aether is filled with vibrations the world is dark. But one day man opens his seeing eye, and there is light.'[1]

APPENDIX I

Some of Professor Heisenberg's very recent reflexions bear on the thesis of the first part of ch. VI; a few remarks may not be amiss.

Dissatisfaction with elementary particle theory has been growing; formal difficulties (divergencies in integrals at large frequencies, etc.) have complicated the conceptual pattern. The new crop of mesons are dealt with in an almost taxonomic way, even though recent work of Yang and Lee on the one hand and of Gell-Mann on the other shows great promise. No theory allows us to calculate, for instance, meson mass and charge ratios, and ingenious attempts to repair the theory—such as the technique of renormalization (Bethe)—lead to breakdown in other places. Thus renormalization (which introduces non-Hermitean operators) ruins the unitary character of the scattering matrices, requires negative probabilities, and invokes strange 'ghost' states which lack a physical interpretation.

All this has made Heisenberg seek a new departure. He argues that we need a fundamental equation for *all* matter, not just a collection of descriptions of particular elementary particles. His reasoning is as in section A (ch. VI): one cannot explain the properties of a class of entities by appealing to entities which possess those properties, that is, which are members of the class. Electrons, protons, neutrons, photons; the wave functions of some one of these particles is usually taken as fundamental to a theory. But each of these is itself a composite, property-possessing particle (even though the properties are unfamiliar in the macrophysical world). Heisenberg's new equation, $\gamma_0 \delta\psi / \delta\psi_v + l^2 \psi(\psi + \psi) = 0$, is meant to be a *generalized* equation for all matter. It accords to matter-in-general none of the properties possessed by individual particles; this move is the only way to account for the spectrum of all the particles we have already observed. Heisenberg gives the credit for this insight to the early Greek philosopher Anaximander, from whose views he claims to have drawn many important ideas.

The new theory apparently meets many objections against the current one. Its scattering matrices are unitary and hence require no negative probabilities; its ghost states are not vicious, and it gives the properties of the photon, in addition to the other particles —an achievement for any elementary particle theory. The details cannot be pursued, but the philosophical reasons behind Heisenberg's dissatisfaction with the current notations give a striking illustration for section A.

APPENDIX II

Wave-packets, pulses of disturbance maxima, etc., have been dealt with in a qualitative way; this has helped in locating large features in the pattern of microphysical inquiry. The 'packets' or 'pulses' are not more than an analogy, instrumental to an exposition of Schrödinger's Ψ function. Let us pursue these points somewhat less informally. The mathematics which follows is meant to mark out the conceptual pattern of the foregoing in the actual symbolism of wave mechanics;[1] it will focus attention on just those features of the theory about which there has been controversy concerning interpretation.

Consider light: the rays are the orthogonal trajectories of the wave surfaces, and, given screens with large apertures, we can ignore diffraction and compute the ray paths by Fermat's Principle

$$\delta \int_{P_0}^{P_1} n\, ds = \delta \int_{P_0}^{P_1} \frac{ds}{u} = 0,$$

where u is the phase velocity.

Refractive index is here a function of position; this is a paradigm case of what we called 'W notation' in section D. (With very small objects (10^{-8} cm.), geometrical optics fails us. Diffraction phenomena can no longer be ignored, nor can they be described in terms of light rays.)

Consider now classical particles. The equation

$$\delta W = \delta \int_{t_0}^{t_1} L\, dt = \delta \int_{t_0}^{t_1} (T - U)\, dt = \delta \int_{t_0}^{t_1} (2T - E)\, dt = 0$$

is the mathematical expression of Hamilton's Principle of Least Action. The natural motion of a system has an extreme value taken between two configurations of the system, when

$$\int_{t_0}^{t_1} L\, dt = \int_{t_0}^{t_1} (T - U)\, dt$$

(the time integral of the Lagrangian function). Here is an element of what we earlier called 'P notation'.

The analogy with optics is clear when comparison orbits are

restricted to those of the same total energy. Thus, for a particle (by Hamilton's Principle)

$$0=\delta\int_{t_0}^{t_1}(2T-E)\,dt=\delta\int_{t_0}^{t_1}2T\,dt=\delta\int_{t_0}^{t_1}mv^2\,dt.$$

Now the length of path is introduced as a variable of integration in place of the time

$$0=\delta\int_{t_0}^{t_1}mv\frac{ds}{dt}dt=\delta\int_{P_0}^{P_1}mv\,ds,$$

or,

$$\delta\int_{P_0}^{P_1}v\,ds=0 \quad \text{(since } m \text{ is constant).}$$

Compare the form of this with our first equation,

$$\delta\int_{P_0}^{P_1}n\,ds=0.$$

If we wanted to form a *PW* notation, associating waves with elementary particles as light waves are associated with light rays, we would have to set the phase velocity u proportional to $1/v$, so that Fermat's Principle holds for matter waves. This was the Schrödinger-De Broglie master-stroke. Thus, if u is the potential energy:

$$v=\sqrt{[2/m(E-U)]}, \quad \text{or} \quad u=\frac{C}{\sqrt{(E-U)}}.$$

A wave function is needed. We introduce a frequency:

$$E=h\gamma, \quad u=\frac{C}{\sqrt{(h\gamma-U)}},$$

so that the phase velocity depends on γ. Thus dispersion exists with these 'waves'. We must now distinguish the phase velocity u and the group velocity u_g. The latter is directly measurable by a recording instrument; the extrema in a wave train of constant amplitude are indistinguishable.[1] On the other hand, with a particle, one can always measure the actual velocity. Hence we require that the group velocity of the waves associated with a particle shall coincide with the measurable velocity of the particle. Thus, making $u_g=v$, we get $\left(\text{by } \frac{1}{V}=\frac{d(\gamma/v)}{d\gamma}\right)$

$$\frac{1}{u_g}=\frac{d(\gamma/u)}{d\gamma}=\frac{1}{\gamma}=\frac{1}{\sqrt{[2/m(E-U)]}}.$$

Thus

$$\frac{\gamma}{u}=\int\frac{d\gamma}{\sqrt{[2/m(h\gamma-U)]}}=\frac{1}{h}\sqrt{[2m(h\gamma-U)]}.$$

Replacing C by $h\gamma/\sqrt{(2m)}$,

$$u=\frac{h\gamma}{\sqrt{[2m(h\gamma-U)]}}.$$

The wave-length of the matter waves is

$$\lambda=\frac{u}{\gamma}=\frac{h}{\sqrt{[2m(h\gamma-U)]}}=\frac{h}{\sqrt{[2m(E-U)]}}.$$

In a region free of force, then,

$$\lambda_0=\frac{h}{\sqrt{(2mE)}}=\frac{h}{mv},$$

which is De Broglie's formula—the conceptual key to the wave-particle duality for matter. The first stage of the calculation is complete. We have been able to localize a wave disturbance in such a way that its dynamical properties are analogous to those of a classical particle, so giving us our *PW* notation. The second stage consists in adapting this for elementary particles.

We require reasons for so combining these concepts: the complex occasion for the De Broglie-Schrödinger retroduction has already been examined. Our reasons may consist simply in interference phenomena in particles, as follows.

If an electron has traversed a difference of potential V, and its De Broglie wave-length is calculated by $\lambda_0=h/mv$, then

$$\lambda_0=\sqrt{(150/V)}$$

(V is in volts and λ in angstrom units). Electrons of several thousand volts have wave-lengths like those of X-rays; hence we expect the sort of interference and diffraction phenomena which obtain for X-rays. As we have seen, this was observed not only with electrons, but also with β-particles, γ-ray photons and even atoms; the process has been repeated and refined hundreds of times. Before these experiments there was not much to choose between wave mechanics and matrix mechanics, save that the former was mathematically more tractable, even if slightly difficult to interpret observationally; Heisenberg's matrices kept close to observation, but they were formally cumbersome. The particle-diffraction experiments shifted microphysics towards the Schrödinger formulation for a time, until it was seen that wave mechanics (treated as a literal description of the electromagnetic field behaviour of individual elementary particles) was a hopeless *cul-de-sac*. The turning point was the transformation theory of Jordan and, more especially, that of Dirac, within which the wave equation is completely dissociated from any classical field interpretation. But this is anticipating: we must return to our wave-mechanical exposition which was developing, as it did for Schrödinger, as an analogy with localized field behaviour—without any quantum discontinuities.

In optics we require a wave differential equation to account for diffraction phenomena—similarly in quantum mechanics. An analogous wave equation turns up:

$$\Delta\Psi = \frac{1}{u^2}\frac{\delta^2\Psi}{dt^2},$$

where

$$u = \frac{h\gamma}{\sqrt{[2m(h\gamma - U)]}} = \frac{E}{\sqrt{[2m(E-U)]}}.$$

The interpretation of Ψ will be dealt with below.

As in classical wave theory, we take a simple periodic function to represent the dependence on time: $\Psi = \psi e^{2\pi i\gamma t} = \psi e^{2\pi i(E/h)t}$. Thus, for the variation of ψ with position we get the fundamental equation of Schrödinger,[1]

$$\Delta\psi + \frac{8\pi^2 m}{h^2}(E-U)\psi = 0.$$

Ψ then, is related to the motion of electrons just as light waves are related to the motion of photons. As the photon is the quantum aspect of its associated field, so too the electron is the quantum aspect of whatever it is that Ψ represents. Hence, Ψ must vanish at infinity when we seek a solution of the wave equation to represent the motion of an electron in an atom. Moreover, any solution (to be useful) must be a continuous, single-valued function of position in a finite region.

These conditions can determine values which, in the quantum mechanics of Bohr, had to be singled out from a range of possible solutions by imposing *ad hoc* conditions. We are on the way here to a more satisfactory *explicans*. It is clear what we have to explain; we must work back from these data to the pattern-concepts.

'Single-valuedness' is equivalent to what Bohr calls 'a sufficiently large periodic orbit'. Moving in an orbit, Ψ must return to its original value—it must 'join up' smoothly—in order to meet the condition that it be single valued. This means that the phase of the wave motion must increase by an integral multiple of 2π. Here is the conceptual condition dealt with intuitively in §C, ch. VI.

Generally, Ψ may vary with position if the index of refraction is variable. The condition of single-valuedness, then, may be expressed:

$$\oint \frac{dq}{\lambda} = \oint \frac{1}{h} \sqrt{[2m(E-U)]}\, dq = n.$$

But

$$\frac{1}{2m}\left(\frac{dS}{dq}\right)^2 + U = E, \quad \text{or} \quad p = dS/dq = \sqrt{[2m(E-U)]}$$

(by the Hamilton-Jacobi), which gives Bohr's quantum condition of sufficiently large periodic orbit[1]

$$\oint p\, dq = nh.$$

The condition of single valuedness can be satisfied only for certain values of E, the eigen-values (i.e. 'proper' or 'characteristic' values) of the differential equation. The physicist's main task in any quantum mechanical problem is to determine eigen-values. These are the selected energy levels.

Originally, Schrödinger regarded Ψ as follows: elementary particles are not localized point charges; rather, their mass and charge are smeared (*verschmiert*) over a certain region. This was my approach, too, in exposition D. But, though effective for bringing out the logic of the uncertainty relations, this approach runs into grave difficulties. Schrödinger took the density of charge to be proportional to $\Psi\overline{\Psi}$ (the square of the amplitude of the function);[2] this is analogous to the role of u_g in our simple wave-particle calculation earlier, but it is much more general. Assume the development of only a single state of the atom (eigen-value E_m). The corresponding Ψ function, then, is

$$\Psi_m = \psi_m e^{2\pi i (E_m/h)t}.$$

ψ_m is real, so, multiplying by Ψ_m, the time factor disappears: thus, the distribution of charge is constant in time. Such a configuration cannot radiate. This is *required* by Bohr's first postulate: of the continuous sequence of mechanically possible orbits, only a discrete set is capable of existence. These orbits are such that no radiation takes place when the electron is in one of them. Bohr got this condition by fiat, but Schrödinger gets it as a consequence of his '*PW*' notation.

ψ functions will now be taken to be orthogonal, and will also be normalized to unity. Since there is an arbitrary multiplicative constant in the solution to the differential equation, we can choose it so as to make the integral of the square of any proper function, ψ_m, equal to unity.

So we specify that all eigen-functions are normalized:

$$\int \psi_m^2 \, d\tau = 1.$$

That ψ is orthogonal entails, by Green's Theorem

$$\int (\psi_m \Delta\psi_n - \psi_n \Delta\psi_m) \, d\tau = \oint (\psi_m \operatorname{grad} \psi_n - \psi_n \operatorname{grad} \psi_m) \, dS,$$

that the integral over the entire region of the product of two proper functions belonging to different eigen-values is zero. Thus:

$$\int \psi_m \psi_n \, d\tau = 0.$$

The eigen-function ψ_m satisfies

$$\Delta\psi_m + \frac{8\pi^2 m_0}{h^2}(E_m - U)\psi_m = 0,$$

and ψ_n satisfies

$$\Delta\psi_n + \frac{8\pi^2 m_0}{h^2}(E_n - U)\psi_n = 0.$$

Multiplying the first equation by ψ_n, the second by ψ_m, and subtracting the first from the second, we have:

$$\psi_m \Delta\psi_n - \psi_n \Delta\psi_m = \frac{8\pi^2 m_0}{h^2}(E_m - E_n)\psi_n \psi_m.$$

Multiply this by *dt* and integrate over all space. The left-hand integral transforms into a surface integral; and, since the eigen-functions vanish exponentially at infinity, this integral is zero. Orthogonality is thus demonstrated.

The wave equation is linear. Hence, any linear combination, e.g.

$$\Psi = c_0\Psi_0 + c_1\Psi_1 + \ldots c_k\Psi_k = c_0\psi_0 e^{2\pi iE_0/ht} + c_1\psi_1 e^{2\pi iE_1/ht} + \ldots$$

is also a solution. For Schrödinger this was an atomic state in which several natural frequencies are simultaneously developed (the amplitude factors c_k being a measure of the excitation). Using this solution to obtain $\Psi\overline{\Psi}$ we get

$$\Psi\overline{\Psi} = c_0^2\psi_0^2 + c_1^2\psi_1^2 + \ldots c_k^2\psi_k^2 + 2c_0c_1\psi_0\psi_1 \cos\frac{2\pi}{h}(E_1 - E_0)t + \ldots$$
$$+ 2c_kc_l\psi_k\psi_l \cos\frac{2\pi}{h}(E_l - E_k)t.$$

Charge density is thus composed of one part constant in time, and another part whose magnitude oscillates with the frequencies

$$\gamma_{kl} = \frac{E_l - E_k}{h}.$$

This variable charge, with the nucleus, represents a variable dipole moment, which, of course, emits light of frequencies γ_{kl}. Averaged in time, the contributions of the variable terms to the charge density vanish. Because the total charge must be constant, the amplitude factors must satisfy

$$\Sigma c_k^2 = 1.$$

The dipole moment $\left(\text{emission from which is calculated according to } \overline{S} = \frac{16\pi^4 c}{3\lambda^4} p_0^2\right)$, then becomes $P_{kl} = -2c_kc_le\int r\psi_k\psi_l d\tau$, where r is the radius vector drawn from the nucleus to the volume element $d\tau$. This automatically fulfills the second postulate of the old quantum mechanics: emission or absorption of light occurs either as the atom passes from a higher to a lower energy state, or vice versa. (Here, Bohr again introduced h without justification.)[1]

The Schrödinger formalism succeeded very well up to this point. It achieved even more[2] before being built into Dirac's transformation theory, but with these further details we are not concerned. The only goal has been to make clear how two basically contradictory physical concepts were harnessed formally into a single system which unified heaps of as-yet-uncorrelated observations. As often happens, however, the formal unifications and retroductions can lead to controversy over the interpretation of the *explicans* —in elementary particle theory this *explicans* is the ψ function. Asking what '$|\psi(x)|^2$' means, will bring us full circle to the situation with which we began, Tycho and Kepler watching the dawn. For $|\psi(x)|^2$ has organized the thinking of microphysicists in very different ways.

For Schrödinger, elementary particles have always been pulses of interfering electromagnetic wave-maxima; Ψ waves are construed as vibrations of the charge of the electron continuously distributed over the wave field. He writes:

> Something exists in the atom which actually vibrates with the observed frequency, viz. a certain part of the electric density-distribution,...the square of the absolute value of ψ is proportional to an electric density, which causes emission of light according to the laws of ordinary electrodynamics...we shall have to postulate $\sum_k c_k^2 = 1$ in order to make the total charge equal to the electronic charge (which we feel inclined to do). ...Is it quite certain that the conception of energy, indispensable as it is in macroscopic phenomena, has *any* other meaning in micromechanical phenomena, than the number of vibrations in h seconds?[1]

Thus Schrödinger's first great step in wave mechanics was—as in the expositions above (sections D, F)—to replace the single point-mass moving in a field of force by the partial differential equations for a wave field.[2] The electron was construed by Schrödinger as a 'charge-cloud', *verschmiert* over space ($\rho = e\psi\overline{\psi}$). Sommerfeld was able to provide a dynamical generalization of this 'fluctuating cloud-charge', the result being a 'mass-cloud' for the electron.[3]

Schrödinger now formed the wave-packet ($|\psi(q)|^2$) as a system whose centre of gravity behaves like a point-mass vibrating harmonically.[4] This charge distribution constitutes a continuous distribution of matter throughout all space; the part of the Ψ field within which the density of charge is sensibly different from zero is, however, only of atomic dimensions. An interpretation of Bohr's frequency condition follows straightway. The older picture of the quantum transition of the electron from one stationary path to another is now replaced by the picture of a partial vibration of energy E_m gradually passing over into one of energy E_n, in which process (during the waning of the one partial vibration and the waxing of the other) the energy difference is radiated out as an electromagnetic wave whose frequency agrees with the beat frequency of the two ψ waves.

This interpretation worked well in single electron problems and with the hydrogen atom; but that it did so was due only to the accidental coincidence in this case of three-dimensional actual space with Schrödinger's n-dimensional phase space. Even a two-body problem is enough to inundate Schrödinger's account with fictitious phase waves and unobservable parameters. The wave amplitude equation in this case is

$$1/m\Delta_1\psi + 1/m\Delta_2\psi + 8\pi^2/h^2(E-U)\psi = 0,$$

which is six dimensional; hence the interpretation is useless in experiments with electrons. It dashed De Broglie's early hopes that field-physics might become reinstated in particle theory. His *ondes de phase* and Schrödinger's treatment of the H-atom quickened the pulses of those physicists who regretted seeing classical physics founder on the quantum reef. Many-electron problems, however, required Ψ waves to be vibrations not in physical space, but in the formal 'configuration space' of the system in question. This space is a manifold of as many dimensions as the system has degrees of freedom. As we have just seen, for two particles it has six dimensions. For four particles it has twelve. In general, given a system of N particles, ψ is a function of $3N$ co-ordinates; its configuration space is always $3N$ dimensional. So although wave mechanics is built on the partial differential equations

$$\Delta\psi + [(8\pi^2\mu)/h^2](E-U)\psi = 0 \quad \text{and} \quad \{H[(h/2\pi i)(\partial/\partial q)\,q] - E\}\,\psi = 0,$$

there can be no question of reviving a field-physics like hydrodynamics, elasticity, or Maxwell's electrodynamics as a conceptual pattern for elementary processes.[1]

Einstein never deserted the hope of this revival, but Bohr, Heisenberg, Dirac and von Neumann would seem to have shown that a classical field-physics for micro-particles is impossible in principle.[2] The 'Schrödinger smear' is particularly inept at explaining collisions between high-energy particles. Heisenberg's system of matrix mechanics avoids these problems (though it has others); it contains nothing corresponding directly with the ψ function. Born[3] construes ψ in a statistical way; $|\psi(q)|^2$ is a 'probability packet'. Like Schrödinger, he clings to a classical picture. Electrons are singularities, point particles. $|\psi(q)|^2$ denotes only the probability of finding a particle in the region of co-ordinates specified in the packet (when we know by what eigenfunctions the system's state may be characterized). The Ψ packet is thus a measure of the chances of finding the particle: if ψ is the wave amplitude of a free electron, and if $\psi=\psi_0$ in the range of positions B to $(B+\Delta B)$, then $\psi_0^2\Delta B$ is the probability that the electron (i.e. the singularity) is in that range. This is provided, of course, that B is multiplied by a suitable constant such that $\psi^2\Delta B$ summed over all B values equals 1, i.e. the total probability of finding the electron is 100%: the matrix must be unitary; there must *be* an electron to find.

A simple example of this interpretation follows from the diffraction experiments of Davisson and Germer, Thomson and Rupp, mentioned in section D. Let the ray intensities diminish so much that only single electrons are passing through the crystal powder at any one instant. Were each electron literally a complete wave, a diffraction pattern with all its concentric rings should appear simultaneously on the screen, similar to when a pebble breaks the surface of a quiet pond. But, in the first place, for electrons this is inconceivable; in the second place, the electrons do in fact strike the screen as point scintillations. In time, a pattern, shown in fig. 28, builds up on the screen; fig. 29 shows this screen viewed on edge, with a distribution intensity scale above it. It is not possible to say where any one electron will appear, but we can say where it is most unlikely to appear, namely between the rings. Thus, $|\psi(x)|^2$ may be construed as the probability of an electron hitting the screen at some given single point.[1] Where the distance from B to ΔB is long, $|\psi(x)|^2$ is small for any particle at any point, but covers a greater range, i.e. the distribution of particles is more even and uniform, making the collection of other information easier. If the distance from B to ΔB is short, $|\psi(x)|^2$ is greater, but covers a very narrow range, i.e. the scintillations will 'pack' at B, making other information difficult to acquire. Here, in yet another form, are the uncertainty relations.

It is now easy to generalize this interpretation for a beam of particles, e.g. a cathode-ray, a γ-ray, a β-ray, and so on. This is most useful, for *these* are the sorts of phenomena with which experimenters must work, not single particles, and certainly not 'electromagnetic' ψ waves. The rays are scattered by diffractors (crystal powders, metal foils, reflector-gratings) into their component particles, and the scattering matrices [the distributions of, e.g. electrons, or γ-ray photons, on the photographic plate which catches them] provide the data for old and for new theories. So now $|\psi(q)|^2$, or better $|\psi(x, y, z; t)|^2$, is the average density of electrons in the beam at the point (x, y, z).[1] This is, for certain

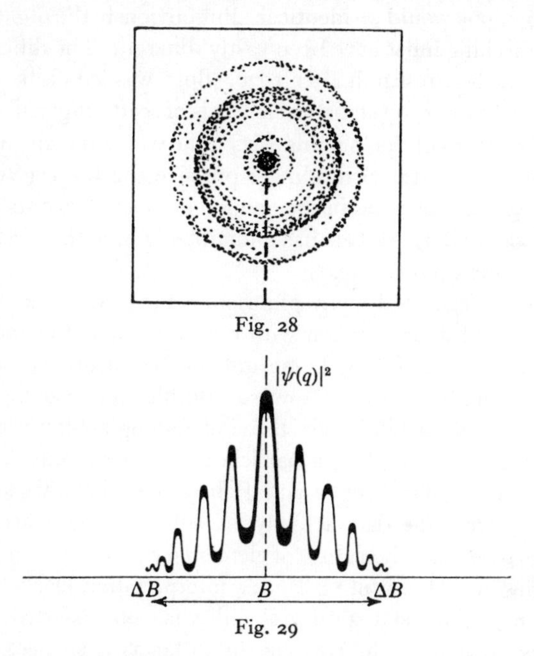

Fig. 28

Fig. 29

purposes, a more useful way of saying what was said in the last paragraph, namely that $(|\psi|^2\, dx\,dy\,dz)$ is the probability of finding an electron at t in the volume element $dx\,dy\,dz$.

Born's point-particle electron appears, in the light of this interpretation, like a return to the second exposition of the uncertainty relations given in section D. There we discussed actual diaphragm experiments as they affected our ability to determine the state of a particle. The initial conclusion of that exposition, however, was erroneous; it was that the uncertainty relations were only a technical limitation in quantum physics, not a limitation in principle.[2] Despite their differences, Born and Schrödinger both accept the uncertainty relations as a *limitation in principle*. But for Schrödinger they describe the inevitable spread of the wave pulse, while for Born they express the impossibility of jointly increasing the probability of locating a particle on a screen and the probability of inferring its momentum.

For most practical microphysical problems, Born and Schrödinger would have made the same theoretical calculations; their solutions and predictions would be identical. But obviously the organization of their thinking must have been vastly different. The difference is reflected in their research after 1930. Born was led on to work on collision behaviour, on statistical analysis of scattering matrices and on the properties of the 'singularities'; the waves remained simply pulses of probability. Schrödinger pursued the *Wellengeist* of the elementary particles; and in the opinion of most physicists this has been fruitless. Born's attack has undoubtedly been the most profitable up to the present day.

Some physicists and many philosophers (e.g. Cassirer, Popper) have expressed dissatisfaction with the development of the statistical interpretation of ψ by Dirac and von Neumann. Einstein has been particularly irreverent towards Born's 'dice-playing God'. For Born the probabilities are those of finding an atom in a particular state, or of finding a particle in a certain condition; but Einstein, Bohm, De Broglie and Jeffreys feel that this leaves in doubt just what the data actually are. Schrödinger's account at least tries to explain the nature of elementary particles; but Jeffreys argues that it is doubtful on Born's interpretation just what proposition it is whose statistical probability is being asserted. Born's emphasis on finding the particle, in so far as it silences further inquiry, veils the nature of the particle itself.[1]

In general, however, such arguments as Jeffreys' are obscure: it may be that their authors are confused about what they would like quantum physics to be. The pronouncements of Einstein[2] and De Broglie[3] are particularly unfortunate. The best case for the 'opposition' is perhaps made by Bohm,[4] whose thesis, about the possibility of hidden variables filling out wave mechanics into a deterministic theory, is an important contribution to philosophy of physics and to discussions of the interpretation of ψ.

Bohm remarks the assumption of all quantum theory, namely that the state of a physical system 'is completely specified by a wave function that determines only the probabilities of actual results that can be obtained in a statistical ensemble of similar experiments'.[1] From this he concludes that

> the uncertainty principle is readily deduced...it becomes a contradiction in terms to ask for a state in which momentum and position are simultaneously and precisely defined....The uncertainty principle is... a necessary consequence of the assumption that the wave function and its probability interpretation provide the most complete possible specification of the state of an individual system....[2]

It is this assumption which Bohm challenges, and in so doing he attacks the foundations of the present system, as he must if uncertainty is to be dodged (see section D). He argues that this assumption '...implies a corresponding unavoidable lack of precision in the very conceptual structure, with the aid of which we can think about and describe the behaviour of the system'.[3] He refuses to accept the renunciation, conjecturing that

> ...from the point of view of macroscopic physics, the co-ordinates and momenta of individual atoms are hidden variables, which in a large scale system manifest themselves only as statistical averages. Perhaps then, our present quantum mechanical averages are simply a manifestation of hidden variables, which have not, however, yet been detected directly.[4]

De Broglie advanced a similar argument in 1926, 1927 and 1930;[5] he was pulverized by Pauli.[6] Madelung[7] made a similar suggestion, but von Neumann's critique seemed to make the future of any such position quite hopeless.[8] But Bohm remains undaunted:

> The present interpretation...involves a real physical limitation on the kinds of theories that we wish to take into consideration...there are no secure experimental or theoretical grounds on which we can base such a choice, because this choice follows from hypotheses that cannot conceivably be subjected to an experimental test...we now have an alternative interpretation....[9]

Bohm then sets out some sketchy mathematical details of his proposals. The most significant feature of his revision of quantum physics consists in this:

> ...in our interpretation, the use of a statistical ensemble is (as in the case of classical statistical mechanics) only a practical necessity, and not a reflection of an inherent imitation on the precision with which it is correct for us to conceive of the variables defining the state of the

system..., our suggested interpretation...provides a much broader conceptual framework than that provided by the usual interpretation [all of whose results]...are obtained...if we [assume]:

(1) That the ψ-field satisfies Schrödinger's equation.
(2) That the particle momentum is restricted to $p = \nabla S_{(x)}$.
(3) That we do not predict or control the precise location of the particle, but have, in practice a statistical ensemble with probability density $P_{(x)} = |\psi(x)|^2$. The use of statistics is, however, not inherent in our conceptual structure, but merely a consequence of our ignorance of the precise initial conditions of the particle.[1]

The limitation, in other words, is a technical one and not a limitation in principle.

Bohm then goes on to examine weaknesses in the present interpretation, exploiting particularly the famous one-electron-two-slit experiment, the Einstein, Podolsky, Rosen experiment, and problems associated with 'the fundamental length' (10^{-13} cm.). He sets out a formalism for describing stationary states, scattering problems, many-body problems, transitions between stationary states, and the photo-electric and Compton effects; he even makes some interesting, if rather obscure, suggestions about an experimental proof of the need for a new interpretation of $|\psi(x)|^2$. In the second of his papers Bohm tries to meet von Neumann's formidable proof of the impossibility of such an interpretation of quantum theory. Whether he succeeds cannot be examined here; in conversation Heisenberg, Oppenheimer, Dirac and Bethe have expressed to me their strongest doubts. Still, other quite serious physicists and philosophers seem unwilling to dismiss Bohm's thesis without giving it a sympathetic hearing.

It is clear, then, that the ultimate interpretation of the ψ function is not yet settled. Some physicists of the first rank embrace Bohm's suggestions, though everyone (including Bohm himself) recognizes that it will be no small task to 'actually develop such modifications in any detail'.[2] Others are unimpressed; the organization of their data remains unaffected by the conceptual pattern Bohm advocates. While perhaps of importance in the long run, this is not now an experimental matter: there is no observation that will settle the issue between Bohm, De Broglie and Einstein, and Heisenberg, Born and Dirac. One day there may be, as Bohm conjectures; but then it will be a vastly different issue. We now regard the differences between

Kepler and Tycho, Galileo and Simplicius, Beeckman and Descartes, as having been settled by observation and reflexion. But these physicists could not, in their time, conceive of any such decisive solution. At the moment we cannot conceive of such an experiment concerning the 'real' nature of $|\psi(x)|^2$; in this we are like physicists of the past. In a way, these differences between physicists, present and past, are like the differences between the observers of the figures in ch. I when they saw different things. Conceptual matters are not finally settled immediately 'the' new observation or idea is suggested: Tycho always had misgivings about Kepler's general position; Descartes was never persuaded by Galileo's later discussions of acceleration; Priestley put a quite different interpretation on Lavoisier's 'disproof' of the existence of phlogiston.[1] Michelson and Morley had little idea of the implications of their own experiments. Stark could not accept von Laue's now orthodox analysis of X-ray diffraction.[2] J. J. Thomson gave an interpretation of the photo-electric and Compton effects which opposed that of Einstein and Compton.[3] Now Halpern,[4] Heisenberg and others disagree vigorously with Bohm's approach. It is in these situations that the expression 'natural philosophy' is most fitting. These men are wrestling with conceptual aspects of physical problems; they are exploring patterns of thinking about matter in a way which will determine what tomorrow's experiments will be, and with what sorts of measurements and observations future laboratory workers will concern themselves. The conceptual basis of De Broglie's doctoral thesis seemed 'philosophical' to his examiners, but today $\lambda = h/mv$ is familiar enough even in undergraduate physics, a household expression in any laboratory. The unsettlement of Einstein, Bohm and De Broglie is regarded as 'merely philosophical' by most physicists today. At this stage it would be venturesome to try finally to settle this matter; nonetheless the issue is a living one in contemporary natural philosophy. Its conceptual significance will be missed by anyone who fails to see how much was at work when physicists of the past disagreed, and missed also by anyone who thinks of the history of physics as just a march of better observations and more accurate experiments. This it surely is; but rarely can a man observe what does not yet exist for him as a conceptual possibility.

NOTES

PAGE 1

1 'Handeln vom Netz, nicht von dem, was das Netz beschreibt', L. Wittgenstein, *Tractatus Logico-Philosophicus* (Harcourt, Brace and Co., New York, 1922), 6. 35.

PAGE 4

1 Wär' nicht das Auge sonnenhaft,
Die Sonne könnt' es nie erblicken;

Goethe, *Zahme Xenien* (Werke, Weimar, 1887–1918), Bk. 3, 1805.

2 Cf. the papers by Baker and Gatonby in *Nature*, 1949–present.

PAGE 5

1 This is not a *merely* conceptual matter, of course. Cf. Wittgenstein, *Philosophical Investigations* (Blackwell, Oxford, 1953), p. 196.

2 (1) G. Berkeley, *Essay Towards a New Theory of Vision* (in *Works*, vol. I (London, T. Nelson, 1948–56)), pp. 51 ff.
(2) James Mill, *Analysis of the Phenomena of the Human Mind* (Longmans, London, 1869), vol. I, p. 97.
(3) J. Sully, *Outlines of Psychology* (Appleton, New York, 1885).
(4) William James, *The Principles of Psychology* (Holt, New York, 1890–1905), vol. II, pp. 4, 78, 80 and 81; vol. I, p. 221.
(5) A. Schopenhauer, *Satz vom Grunde* (in *Sämmtliche Werke*, Leipzig, 1888), ch. IV.
(6) H. Spencer, *The Principles of Psychology* (Appleton, New York, 1897), vol. IV, chs. IX, X.
(7) E. von Hartmann, *Philosophy of the Unconscious* (K. Paul, London, 1931), B, chs. VII, VIII.
(8) W. M. Wundt, *Vorlesungen über die Menschen und Thierseele* (Voss, Hamburg, 1892), IV, XIII.
(9) H. L. F. von Helmholtz, *Handbuch der Physiologischen Optik* (Leipzig, 1867), pp. 430, 447.
(10) A. Binet, *La psychologie du raisonnement, recherches expérimentales par l'hypnotisme* (Alcan, Paris, 1886), chs. III, V.
(11) J. Grote, *Exploratio Philosophica* (Cambridge, 1900), vol. II, pp. 201 ff.
(12) B. Russell, in *Mind* (1913), p. 76. *Mysticism and Logic* (Longmans, New York, 1918), p. 209. *The Problems of Philosophy* (Holt, New York, 1912), pp. 73, 92, 179, 203.
(13) Dawes Hicks, *Arist. Soc. Sup.* vol. II (1919), pp. 176–8.

(14) G. F. Stout, *A Manual of Psychology* (Clive, London, 1907, 2nd ed.), vol. II, I and 2, pp. 324, 561–4.
(15) A. C. Ewing, *Fundamental Questions of Philosophy* (New York, 1951), pp. 45ff.
(16) G. W. Cunningham, *Problems of Philosophy* (Holt, New York, 1924), pp. 96–7.

3 Galileo, *Dialogue Concerning the Two Chief World Systems* (California, 1953), 'The First Day', p. 33.

PAGE 6

1 '"Das ist doch kein Sehen!"—"Das ist doch ein Sehen!" Beide müssen sich begrifflich rechtfertigen lassen' (Wittgenstein, *Phil. Inv.* p. 203).

2 Brain, *Recent Advances in Neurology* (with Strauss) (London, 1929), p. 88. Compare Helmholtz: 'The sensations are signs to our consciousness, and it is the task of our intelligence to learn to understand their meaning' (*Handbuch der Physiologischen Optik* (Leipzig, 1867), vol. III, p. 433).

See also Husserl, 'Ideen zu einer Reinen Phaenomenologie', in *Jahrbuch für Philosophie*, vol. I (1913), pp. 75, 79, and Wagner's *Handwörterbuch der Physiologie*, vol. III, section I (1846), p. 183.

3 Mann, *The Science of Seeing* (London, 1949), pp. 48–9. Arber, *The Mind and the Eye* (Cambridge, 1954). Compare Müller: 'In any field of vision, the retina sees only itself in its spatial extension during a state of affection. It perceives itself as... etc.' (*Zur vergleichenden Physiologie des Gesichtesinnes des Menschen und der Thiere* (Leipzig, 1826), p. 54).

4 Kolin: 'An astigmatic eye when looking at millimeter paper can accommodate to see sharply either the vertical lines or the horizontal lines' (*Physics* (New York, 1950), pp. 570ff.).

5 Cf. Whewell, *Philosophy of Discovery* (London, 1860), 'The Paradoxes of Vision'.

6 Cf. e.g. J. Z. Young, *Doubt and Certainty in Science* (Oxford, 1951, The Reith Lectures), and Gray Walter's article in *Aspects of Form*, ed. by L. L. Whyte (London, 1953). Compare Newton: 'Do not the Rays of Light in falling upon the bottom of the Eye excite Vibrations in the Tunica Retina? Which Vibrations, being propagated along the solid Fibres of the Nerves into the Brain, cause the Sense of seeing' (*Opticks* (London, 1769), Bk. III, part I).

PAGE 7

1 'Rot und grün kann ich nur sehen, aber nicht hören' (Wittgenstein, *Phil. Inv.* p. 209).

2 Cf. 'An appearance is the same whenever the same eye is affected in the same way' (Lambert, *Photometria* (Berlin, 1760)); 'We are justified, when different perceptions offer themselves to us, to infer that the underlying real conditions are different' (Helmholtz, *Wissenschaftliche Abhandlungen* (Leipzig, 1882), vol. II, p. 656), and Hertz: 'We form for ourselves images or symbols of the external objects; the manner in which we form them is such that the logically necessary (*denknotwendigen*) consequences of the images in thought are invariably the images of materially necessary (*naturnotwendigen*) consequences of the corresponding objects' (*Principles of Mechanics* (London, 1889), p. 1).

Broad and Price make depth a feature of the private visual pattern. However, Weyl (*Philosophy of Mathematics and Natural Science* (Princeton, 1949), p. 125) notes that a single eye perceives qualities spread out in a *two*-dimensional field, since the latter is dissected by any one-dimensional line running through it. But our conceptual difficulties remain even when Kepler and Tycho keep one eye closed.

Whether or not two observers are having the same visual sense-data reduces directly to the question of whether accurate pictures of the contents of their visual fields are identical, or differ in some detail. We can then discuss the publicly observable pictures which Tycho and Kepler draw of what they see, instead of those private, mysterious entities locked in their visual consciousness. The accurate picture and the sense-datum must be identical; how could they differ?

PAGE 8

1 From the B.B.C. report, 30 June 1954.

2 Newton, *Opticks*, Bk. II, part I. The writings of Claudius Ptolemy sometimes read like a phenomenalist's textbook. Cf. e.g. *The Almagest* (Venice, 1515), VI, section 11, 'On the Directions in the Eclipses', 'When it touches the shadow's circle from within', 'When the circles touch each other from without'. Cf. also VII and VIII, IX (section 4). Ptolemy continually seeks to chart and predict 'the appearances'—the points of light on the celestial globe. *The Almagest* abandons any attempt to explain the machinery behind these appearances.

Cf. Pappus: 'The (circle) dividing the milk-white portion which owes its colour to the sun, and the portion which has the ashen colour natural to the moon itself is indistinguishable from a great circle' (*Mathematical Collection* (Hultsch, Berlin and Leipzig, 1864), pp. 554–60).

3 This famous illusion dates from 1832, when L. A. Necker, the Swiss naturalist, wrote a letter to Sir David Brewster describing how when certain rhomboidal crystals were viewed on end the perspective could shift in the way now familiar to us. Cf. *Phil. Mag.* III, no. 1 (1832),

329–37, especially p. 336. It is important to the present argument to note that this observational phenomenon began life not as a psychologist's trick, but at the very frontiers of observational science.

4 Wittgenstein answers: 'Denn wir sehen eben wirklich zwei verschiedene Tatsachen' (*Tractatus*, 5. 5423).

PAGE 9

1 'Auf welche Vorgänge spiele ich an?' (Wittgenstein, *Phil. Inv.* p. 214).

PAGE 10

1 *Ibid.* p. 194 (top).

PAGE 11

1 *Ibid.* p. 200.

2 This is *not* due to eye movements, or to local retinal fatigue. Cf. Flugel, *Brit. J. Psychol.* VI (1913), 60; *Brit. J. Psychol.* V (1913), 357. Cf. Donahue and Griffiths, *Amer. J. Psychol.* (1931), and Luckiesh, *Visual Illusions and their Applications* (London, 1922). Cf. also Peirce, *Collected Papers* (Harvard, 1931), 5, 183. References to psychology should not be misunderstood; but as one's acquaintance with the psychology of perception deepens, the character of the conceptual problems one regards as significant will deepen accordingly. Cf. Wittgenstein, *Phil. Inv.* p. 206 (top). Again, p. 193: 'Its causes are of interest to psychologists. We are interested in the concept and its place among the concepts of experience.'

3 Wittgenstein, *Phil. Inv.* p. 212.

4 From Boring, *Amer. J. Psychol.* XLII (1930), 444 and cf. Allport, *Brit. J. Psychol.* XXI (1930), 133; Leeper, *J. Genet. Psychol.* XLVI (1935), 41; Street, *Gestalt Completion Test* (Columbia Univ., 1931); Dees and Grindley, *Brit. J. Psychol.* (1947).

PAGE 12

1 Köhler, *Gestalt Psychology* (London, 1929). Cf. his *Dynamics in Psychology* (London, 1939).

2 'Mein Gesichteseindruck hat sich geändert;—wie war er früher; wie ist er jetzt?—Stelle ich ihn durch eine genaue Kopie dar—und ist das keine gute Darstellung?—so zeigt sich keine Änderung' (Wittgenstein, *Phil. Inv.* p. 196).

3 'Was gezeigt werden kann, kann nicht gesagt werden' (Wittgenstein, *Tractatus*, 4. 1212).

PAGE 13

1 This case is different from fig. 1. Now I can help a 'slow' percipient by tracing in the outline of the bear. In fig. 1 a percipient either gets the perspectival arrangement, or he does not, though even here Wittgenstein makes some suggestions as to how one might help; cf. *Tractatus*, 5. 5423, last line.

2 Wittgenstein, *Phil. Inv.* p. 193. Helmholtz speaks of the 'integrating' function which converts the figure into the appearance of an object hit by a visual ray (*Phys. Optik*, vol. III, p. 239). This is reminiscent of Aristotle, for whom seeing consisted in emanations from our eyes. They reach out, tentacle-fashion, and touch objects whose shapes are 'felt' in the eye. (Cf. *De Caelo* (Oxford, 1928), 290a, 18; and *Meteorologica* (Oxford, 1928), III, iv, 373b, 2. (Also Plato, *Meno*, London, 1869), 76C–D.) But he controverts this in *Topica* (Oxford, 1928), 105b, 6.) Theophrastus argues that 'Vision is due to the gleaming...which [in the eye] reflects to the object' (*On the Senses*, 26, trans. G. M. Stratton). Hero writes: 'Rays proceeding from our eyes are reflected by mirrors...that our sight is directed in straight lines proceeding from the organ of vision may be substantiated as follows' (*Catoptrics*, 1–5, trans. Schmidt in *Heronis Alexandrini Opera* (Leipzig, 1899–1919)). Galen is of the same opinion. So too is Leonardo: 'The eye sends its image to the object...the power of vision extends by means of the visual rays...' (*Notebooks*, C.A. 135 v.b. and 270 v.c.). Similarly Donne in *The Ecstasy* writes: 'Our eye-beams twisted and...pictures in our eyes to get was all *our* propagation.'

This is the view that all perception is really a species of touching, e.g. Descartes' *impressions*, and the analogy of the wax. Compare: '[Democritus] explains [vision] by the air between the eye and the object [being] compressed...[it] thus becomes imprinted..."as if one were to take a mould in wax"...' Theophrastus (*op. cit.* 50–3). Though it lacks physical and physiological support, the view is attractive in cases where lines seem suddenly to be forced into an intelligible pattern—by us.

PAGE 14

1 *Ibid.* p. 193. Cf. Helmholtz, *Phys. Optik*, vol. III, pp. 4, 18 and Fichte (*Bestimmung des Menschen*, ed. Medicus (Bonn, 1834), vol. III, p. 326). Cf. also Wittgenstein, *Tractatus*, 2. 0123.

2 P. B. Porter, *Amer. J. Psychol.* LXVII (1954), 550.

PAGE 15

1 Writings by Gestalt psychologists on 'set' and 'Aufgabe' are many. Yet they are overlooked by most philosophers. A few fundamental

papers are: Külpe, *Ber. I Kongress Exp. Psychol., Giessen* (1904); Bartlett, *Brit. J. Psychol.* VIII (1916), 222; George, *Amer. J. Psychol.* XXVIII (1917), 1; Fernberger, *Psychol. Monogr.* XXVI (1919), 6; Zigler, *Amer. J. Psychol.* XXXI (1920), 273; Boring, *Amer. J. Psychol.* XXXV (1924), 301; Wilcocks, *Amer. J. Psychol.* XXXVI (1925), 324; Gilliland, *Psychol. Bull.* XXIV (1927), 622; Gottschaldt, *Psychol. Forsch.* XII (1929), 1; Boring, *Amer. J. Psychol.* XLII (1930), 444; Street, *Gestalt Completion Test* (Columbia University, 1931); Ross and Schilder, *J. Gen. Psychol.* X (1934), 152; Hunt, *Amer. J. Psychol.* XLVII (1935), 1; Süpola, *Psychol. Monogr.* XLVI (1935), 210, 27; Gibson, *Psychol. Bull.* XXXVIII (1941), 781; Henle, *J. Exp. Psychol.* XXX (1942), 1; Luchins, *J. Soc. Psychol.* XXI (1945), 257; Wertheimer, *Productive Thinking* (1945); Russell Davis and Sinha, *Quart. J. Exp. Psychol.* (1950); Hall, *Quart. J. Exp. Psychol.* II (1950), 153.

Philosophy has no concern with fact, only with conceptual matters (cf. Wittgenstein, *Tractatus*, 4. 111); but discussions of perception could not but be improved by the reading of these twenty papers.

PAGE 16

1 Often 'What do you see?' only poses the question 'Can you identify the object before you?'. This is calculated more to test one's knowledge than one's eyesight.

PAGE 17

1 Duhem, *La théorie physique* (Paris, 1914), p. 218.

2 Chinese poets felt the significance of 'negative features' like the hollow of a clay vessel or the central vacancy of the hub of a wheel (cf. Waley, *Three Ways of Thought in Ancient China* (London, 1939), p. 155).

3 Infants are indiscriminate; they take in spaces, relations, objects and events as being of equal value. They still must learn to organize their visual attention. The camera-clarity of their visual reactions is not by itself sufficient to differentiate elements in their visual fields. Contrast Mr W. H. Auden who recently said of the poet that he is 'bombarded by a stream of varied sensations which would drive him mad if he took them all in. It is impossible to guess how much energy we have to spend every day in not-seeing, not-hearing, not-smelling, not-reacting.'

4 Cf. 'He was blind to the *expression* of a face. Would his eyesight on that account be defective?' (Wittgenstein, *Phil. Inv.* p. 210) and 'Because they seeing see not; and hearing they hear not, neither do they understand' (Matt. xiii. 10–13).

5 'Es hört doch jeder nur, was er versteht' (Goethe, *Maxims* (*Werke*, Weimar, 1887–1918)).

6 Against this Professor H. H. Price has argued: 'Surely it appears to both of them to be rising, to be moving upwards, across the horizon ...they both see a moving sun: they both see a round bright body which appears to be rising.' Philip Frank retorts: 'Our sense observation shows only that in the morning the distance between horizon and sun is increasing, but it does not tell us whether the sun is ascending or the horizon is descending...' (*Modern Science and its Philosophy* (Harvard, 1949), p. 231). Precisely. For Galileo and Kepler the horizon drops; for Simplicius and Tycho the sun rises. This is the difference Price misses, and which is central to this essay.

PAGE 19

1 This parallels the too-easy epistemological doctrine that all normal observers see the same things in *x*, but interpret them differently.

2 Cf. the important paper by Carmichael, Hogan and Walter, 'An Experimental Study of the Effect of Language on the Reproduction of Visually Perceived Form', *J. Exp. Psychol.* XV (1932), 73–86. (Cf. also Wulf, *Beiträge zur Psychologie der Gestalt.* VI. 'Über die Veränderung von Vorstellungen (Gedächtnis und Gestalt).' *Psychol. Forsch.* I (1921), 333–73.) Cf. also Wittgenstein, *Tractatus*, 5. 6; 5. 61.

3 Wittgenstein, *Phil. Inv.* p. 206.

4 '"Seeing as..." is not part of perception. And for that reason it is like seeing and again not like' (*ibid.* p. 197).

5 'All seeing is seeing as...if a person sees something at all it must look like something to him...' (G. N. A. Vesey, 'Seeing and Seeing As', *Proc. Aristotelian Soc.* (1956), p. 114.)

PAGE 21

1 'Is the pinning on of a medal merely the pinning on of a bit of metal?' (Wisdom, 'Gods', *Proc. Aristotelian Soc.* (1944–5)).

PAGE 22

1 Wittgenstein, *Phil. Inv.* p. 212. Cf. *Tractatus* 2. 0121. Cf. Also Helmholtz, *Phys. Optik*, vol. III, p. 18.

PAGE 23

1 Drawn on grid paper the two visual pictures could be geometrically identical. Cf. 'If the two different 'appearances' of a reversible figure were indeed things ('pictures') we could conceive of them projected out from our minds, on to a screen, side by side, and distinguishable. But the only images on a screen which could serve

as projections of the two different 'appearances' would be identical' (G. N. A. Vesey, 'Seeing and Seeing As', *Proc. Aristotelian Soc.* (1956)).

PAGE 24

1 Wittgenstein, *Phil. Inv.* p. 197.

2 'Of the sense-data we cannot know more...than that they are in agreement with one another' (Leibniz, *Die Philosophische Schriften* (Berlin, 1875–90), vol. IV, p. 356).

PAGE 25

1 We speak of 'phototropism' in flatworms, but not seeing. (If dogs talked, Descartes might not have regarded them as blind machines.)

2 In their logical-construction, 'picture-theory' periods, Russell (*Logical Atomism* (Minnesota, 1950)), Wittgenstein (*Tractatus*) and Wisdom ('Logical Constructions' (*Mind*, 1931–4)) must fall into this class.

PAGE 26

1 '"Knowing" it means only: being able to describe it' (Wittgenstein, *Phil. Inv.* p. 185).

2 'I looked at the flower, but was thinking of something else and was not conscious of its colour...[I] looked at it without seeing it...' (*ibid.* p. 211). The history of physics supplies further examples, cf. chs. II, IV and VI.

3 Cf. Kant: 'Intuition without concepts is blind....Concepts without intuition are empty.' Indeed how is 'interpretation' of a *pure* visual sense-datum possible?

PAGE 27

1 Cf. Wittgenstein, *Tractatus*, 2. 1–2, 2 and 3–3. 1.

PAGE 29

1 *Ibid.* 4. 016.

2 Thus the pattern of a picture is not another element of the picture. The difference between the bird and the antelope is like that between *bRt* and *tRb*. We may see different things in the same visual elements; just as when you say '*bRt*' and I say '*tRb*' we have said different things with the same words. Cf. Wittgenstein, *Tractatus*, 3. 141: 'Der Satz ist kein Wörtengemisch.—(Wie das musikalische Thema kein Gemisch von Tönen.)'

PAGE 30

1 '"Natural Philosophy"...lies not in discovering facts, but in discovering new ways of thinking about them. The test which we apply

to these ideas is this—do they enable us to fit the facts to each other?' Bragg, 'The Atom', in *The History of Science* (London, 1948), p. 167.

'Orderly arrangement is the task of the scientist. A science is built out of facts just as a house is built out of bricks. But a mere collection of facts cannot be called a science any more than a pile of bricks can be called a house' (Poincaré, *Foundations of Science* (Science Press, Lancaster, Pa., 1946), p. 127). 'An object is frequently not seen *from not knowing how to see it*, rather than from any defect in the organ of vision...[Herschel said] "I will prepare the apparatus, and put you in such a position that [Fraunhofer's dark lines] shall be visible, and yet you shall look for them and not find them: after which, while you remain in the same position, I will instruct you *how to see them*, and you shall see them, and not merely wonder you did not see them before, but you shall find it impossible to look at the spectrum without seeing them"' (Babbage, *The Decline of Science in England* (R. Clay, London, 1830)).

PAGE 31

1 Cf. Mach, *Mechanics:* 'Apprehension of facts' (p. 5), 'A process of adaptation of thoughts to facts' (p. 7), and 'We err when we expect more enlightenment from an hypothesis than from the facts themselves' (p. 600). J. S. Mill sees facts, collects them, locates them, dates them: see *A System of Logic* (London, 1875), Bk. III.

2 Our concern is not with words. If a philosopher's words did not reveal his conceptions, they would be of no interest. 'The true meaning of a term is to be found by observing what a man does with it, not what he says about it' (Bridgman, *The Logic of Modern Physics* (New York, 1927)).

PAGE 33

1 Wittgenstein, *Phil. Inv.* p. 188.

Cf. 'Eskimo languages contain separate words for "snow on the ground", "snow heap", "falling snow", "drifting snow", "soft snow", etc....that we have but one word for all these varieties of snow may be the reason why we often fail to perceive slight differences in snow which would be immediately obvious to Eskimos....Conversely, language distinctions which seem inevitable to us may be ignored in languages which reflect a very different type of culture...if we spoke a different language, we would perceive a somewhat different world.' Crafts, Schneirla, Robinson and Gilbert, *Recent Experiments in Psychology* (New York and London, 1950), ch. XXII, 'Factors Influencing Perception and Memory'. Cf. also Gibson *J. Exp. Psychol.* XII (1929), 1–39; Seeleman, *Arch. Psychol.* (1940), p. 258. Haggard and Rose, *J. Exp. Psychol.* XXXIV (1944), 45. Luchins, *J. Soc. Psychol.* XXI (1945), 257.

2 *Critique of Practical Reason* (Abbott trans.), p. 150.

3 'Wir können also auch nicht sagen, was wir nicht denken können' (Wittgenstein, *Tractatus*, 5. 61).

PAGE 34

1 Cf. Wittgenstein, *Tractatus*, 3. 032: 'Etwas "der Logik widersprechendes" in der Sprache darstellen, kann...ebensowenig.'

PAGE 36

1 In certain cases, notational limitations can even rule out notions as *logically* impossible. These will be of interest later when logical features of formalized quantum theories are considered.

2 What follows is a development of an argument by A. Koyré, 'La Loi de la Chute des Corps', *Études Galiléennes*, *Histoire de la Pensée* (Hermann and Cie, Paris, 1939). Koyré, however, is not alert to the distinction above—that between the practical limitations of forming a concept '*x*' in a language in which *x* is not easily expressed, and the logical limitations of forming *x* in a language whose structure explicitly forbids the formation of *x*. Ultimately we shall be concerned with the latter. Now the former has our attention. Koyré's otherwise excellent monograph slides from the one possibility to the other without warning.

PAGE 37

1 'Galileo a Paolo Sarpi in Venezia, Padova, 16, ottobre 1604' (*Opere* (Milano, 1953), vol. X, p. 115).

2 For the history of these special formulae see P. Duhem, *Études sur Léonard de Vinci* (Paris, 1913), vol. VIII, and E. Borchert, 'Die Lehre von der Bewegung bei N. Oresme' (*Beiträge zur Geschichte der Philosophie und Theologie des Mittelalters*, Bd. XXXI (Münster, 1934), fasc. 3).

3 'It does not seem expedient to investigate what may be the cause of acceleration' (*Opere*, vol. VII, p. 202).

PAGE 38

1 *La nouva scientia inventa da Nicolo Tartaglia*, I, prop. I.

2 'Tanto major sit semper impressio, quanto magis movetur naturaliter corpus, et continuo novum impetum recipit, cum in se motus causam contineat, quae est inclinatio ad locum suum eundi, extra quem per vim consistit' (J. B. Benedetti, *Diversarum speculationum mathematicarum et physicarum liber* (Taurini, 1585), p. 184).

PAGE 39

1 For more on the theory of impetus, see E. Wohlwill, 'Die Entdeckung des Beharrungsgesetzes', *Zeitschrift für Völkerpsychologie und Sprachwissenschaft*, vols. XIV and XV; P. Duhem, 'De l'accélération produite par une force constante', *Congrès international de Philosophie*, 2nd Session (Geneva, 1905).

2 Alexandri Piccolominei, *In mechanicas questiones Aristotelis paraphrasis paulo quidem plenior* (Rome, 1547), ch. 38. Julii Cesarii Scaligeri, *Exotericarum exercitationum liber XV, De Subtilitate ad Hieronimum Cardanum* (Lutetiae apud Vascosanum), 1557.

3 'La gravité qui descend librc acquiert à chaque degré de temps un degré de mouvement et, à chaque degré de mouvement, un degré de vitesse' (Duhem's translation), *Les Manuscrits de Léonard de Vinci* (Paris, Ravaisson-Mollien, MS. de la Bibliothéque de l'Institut), fol. 44.

4 Duhem points this out: 'De l'accélération', p. 872.

5 Koyré, 'La Loi de la Chute des Corps', p. 88.

PAGE 40

1 Cf. Galileo's own account, given in an exchange between Sagredo and Salviati (*Opere*, vol. VIII, p. 203).

PAGE 41

1 Cf. ch. IV.

2 Cf. ch. IV.

3 Newton still uses the term eighty years later.

4 Duhem, 'De l'accélération', p. 888.

PAGE 43

1 Galileo Galilei, *Frammenti attenenti ai Discorsi*, etc. (*Opere*, vol. VIII, p. 373).

2 Galileo's argument contains a double error: this enables him to reason from the statement that the velocities are proportional to the spaces, ($v \propto s$), which is false, to the true statement: the spaces traversed are

proportional to the squares of the times ($s \propto t^2$). Cf. Duhem, *Études sur Léonard de Vinci* (Paris, 1906–13), vol. III, p. 570ff. In 1638, however (*Two New Sciences* (Macmillan, New York, 1914), p. 168), Galileo proves that v cannot be proportional to d. He confuses here average velocity with instantaneous velocity; 'sum' *qua* geometrical area, is not distinguished from 'sum' *qua* the result of integration. This results from attempting with geometry alone what requires an integral calculus. But the 'proof' is carried out in the same geometrical manner as his reflexions of 1604. [Thus he argues that if $v \propto d$ then the distances must be traversed in equal times. If the velocity for 8′ (2s) were double that for 4′ (s) then the time intervals for both falls would be equal. It would take no longer to fall 2s than to fall s. This means that the distance ($2s-s$) would be traversed instantaneously, *reductio ad absurdum*. So velocities cannot be proportional to distances.] It cannot be, then, that the geometrical notation made it impossible to see both that (1) $v \propto s$ does not entail $s = at^2/2$, and (2) that v cannot be proportional to s; but it made it very, very difficult.

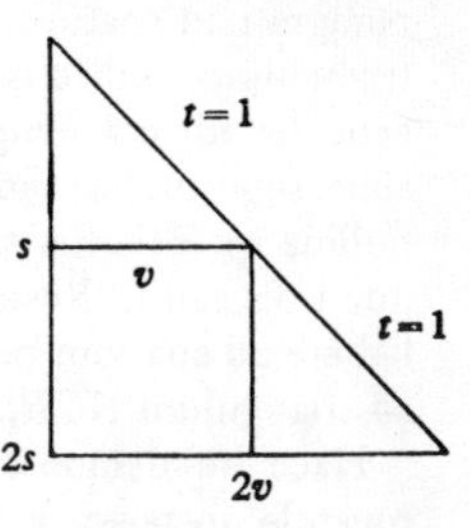

3 Cf. *Journal de Beeckman*, in Descartes' *Oeuvres* (Paris, 1824–6), vol. X, p. 331.

PAGE 44

1 'Mota semel nunquam quiescunt, nisi impediantur. Omnis res semel mota nunquam quiescit, nisi propter externum impedimentum. Quoque impedimentum est imbecillius, eo diutius mota movetur: si enim aliquid in altum projiciatur simulque circulariter moveatur, ad sensum non quiescet ante reditum in terram; et si quiescat tandem id non fit propter impedimentum aequabile, sed propter impedimentum inaequabile, quia alia atque alia pars aeris vicissim rem motam tangit' (*op. cit.* p. 60, note f). This is not the law of inertia, as Duhem ('De l'accélération', p. 904) supposes. It lacks the essential 'in a straight line'. Beeckman, like Hobbes and the younger Galileo, advocated persistance of circular motion only. (The tradition behind this is discussed in ch. IV.) Descartes himself first formulated the law of inertia.

2 Cf. Descartes and Beeckman, 'Physico-mathematica', *Oeuvres*, vol. X, pp. 75ff.

PAGE 45

1 'Cum autem momenta haec sint individua, habebit spacium per quod res una hora cadit *ADE*. Spatium per quod duabus horis cadit,

duplicat proportionem temporis, id est *ADE* ad *ACB*, quae est duplicata proportio *AD* ad *AC*. Sit enim momentum spatij per quod res una hora cadit alicujus magnitudinis, videlicet *ADEF*. Duabis horis perficiet talia tria momenta, scilicet *AFEGBHCD*. Sed *AFED* constat ex *ADE* cum *AFE*; atque *AFEGBHCD* constat ex *ACB* cum *AFE* et *EGB* id est cum duplo *AFE*.

Sic si momentum sit *AIRS*, erit proportio spatii ad spatium, ut *ADE* cum *klmn*, ad *ACB* cum *klmnopqt*, id est etiam duplum *klmn*. Ast *klmn* est multo minus quam *AFE*. Cum igitur proportio spatii peragrati ad spatium peragratum constet ex proportione trianguli ad triangulum, adjectis utrique termino aequalibus, cumque haec aequalia adjecta semper eo minora fiant quo momenta spatii minora sunt: sequitur haec adjecta nullius quantitatus fore quando momentum nullius quantitatus statuitur. Tale autem momentum est spatii per quod res cadit. Restat igitur spatium per quod res cadit una hora se habere ad spatium per quod cadit duabus horis, ut triangulum *ADE* ad triangulum *ACB*.

Haec ita demonstravit M. Perron, cum ei ansam praebuissem, rogando an possit quis scire quantum spatium res cadendo conficeret unica hora, cum scitur quantum conficiat duabus horis, secundum mea fundamenta, viz. *quod semel movetur*, *semper movetur*, *in vacuo* et supponendo inter terram et lapidem cadentem esse vacuum. Si igitur experientia compertum sit, lapidem cedidisse duabus horis per mille pedes, continebit triangulum *ABC* 1000 pèdes. Hujus radix est 100 pro linea *AC* quae respondit horis duabus. Bisecata ea in *D*, respondet *AD* uni horae. Ut igitur se habet proportio *AC* ad *AD* duplicata, id est 4 ad 1, sic 1000 ad 250, id est *ACB* ad *ADE*' (*Journal de Beeckman*, vol. x, p. 58).

2 'Contigit mihi ante paucos dies familiaritate uti ingeniosissimi viri, qui talem mihi quaestionem proposuit: Lapis, aiebat, descendit ab *A* ad *B* una hora; attrahitur autem a terra perpetuo eadem vit, nec quid deperdit ab illa celeritate quae illi impressa est priori attractione. Quod enim in vacuo movetur semper moveri existimabat. Queritur quo tempore tale spatium percurrat' ('Cogitationes Privatae', *Oeuvres*, vol. x, p. 219ff.).

3 'Solvi quaestionem. In triangulo isoscelo rectangulo, *ABC* spatium [motum] repraesentat; inaequalitas spatii a puncto *A* ad basim *BC*, motus inaequalitatem. Igitur *AD* percurritur tempore, quod *ADE* repraesentat; *DB* vero tempore quod *DEBC* repraesentat: ubi est notandum minus spatium tardiorem motum repraesentare. Est autem *AED* tertia pars *DEBC*: ergo triplo tarduis percurret *AD* quam *DB*. Aliter autem proponi potest haec quaestio, ita ut semper vis attractiva terrae aequalis sit illi quae primo momento fuit: nova producitur,

priori remanente. Tunc quaestio solvetur in pyramide' ('Cogitationes Privatae', *Oeuvres*, vol. x, p. 219).

PAGE 46

1 That is, the triangle becomes a pyramid—*solvetur in pyramide.*

2 Both Beeckman and the later Galileo represent change in time by a vertical line and not by the horizontal one familiar to their contemporaries, and to us.

PAGE 48

1 'In proposita quaestione, ubi imaginatur singulis temporibus novam addi vim qua corpus grave tendat deorsum, dico vim illam eodem pacto augeri, quo augentur lineae transversae *DE*, *FG*, *HI*, et aliae infinitae transversae quae inter illas possunt imaginari. Quod ut demonstrem, assumam pro primo minimo vel puncto motus, quod causatur a primo quae imaginari potest attractiva vi terrae, quadratum *ALDE*. Pro secundo minimo motus, habebimus duplum, nempe *DMGF*; pergit enim ea vis quae erat in primo minimo, et alia nova accedit illi aequalis, Item in tertio minimo motus, erunt 3 vires; nempe primi, secundi et tertii minimi temporis, etc. Hic autem numerus est triangularis, ut alias forte fusius explicabo, et apparet hunc figuram triangularem *ABC* repraesentare. Immo, inquies, sunt partes protuberantes *ALE*, *EMG*, *GOI*, etc. quae extra trianguli figuram exeunt. Ergo figura triangulari illa progressio non debet explicari. Sed respondeo illas partes protuberantes oriri ex eo quod latitudinem dederimus minimis, quae indivisibilia debent imaginari et nullis partibus constantia. Quod ita demonstratur. Dividam illud minimum *AD* in duo aequalia in *Q*; iamque *ARSQ* est [primum] minimum motus, et *QTED* secundum minimum motus, in quo erunt duo minima virium. Eodem pacto dividamus *DF*, *FH*, etc. Tunc habebimus partes protuberantes *ARS*, *STE*, etc. Minores sunt parte protuberante *ALE*, ut patet. Rursum, si pro minimo assumam minorem, ut $A\alpha$, partes protuberantes erunt adhuc minores, ut $\alpha\beta\gamma$, etc. Quod si denique pro illo minimo assumam verum minimum, nempe punctum, tum illae partes protuberantes nullae erunt, quia non possunt esse totum punctum, ut patet, sed tantum media pars minimi *ALDE*, atqui puncti media pars nulla est. Ex quibus patet, si imaginetur, verbi gratia lapis ex *A* ad *B* trahi a terra in vacuo per vim quae aequaliter ab illa semper fluat, priori remanente, motum primum in *A* se habere ad ultimum qui est in *B*, ut punctum *A* se habet ad lineam *BC*. Mediam vero partem *GB* triplo celerius pertransiri a lapide, quam alia media pars *AG*, quia triplo majori vi a terra trahitur: spatium enim *FGBC* triplum est spatii *AFG*, ut facile probatur. Et sic proportione dicendum de caeteris partibus' ('Physico-

mathematica', *Oeuvres*, vol. x, pp. 75 ff.). Descartes does employ 'Archimedean' integration. But he nonetheless concludes that $d \propto t^2$ from the principle $v \propto d$. Beeckman's 'mistake' consisted in construing the principle as $v \propto t$.

2 He need not have tumbled. Gravitational attraction between bodies varies inversely as the square of the distance between them, but Newton needed no impetus in this ancient sense to make his theory work.

3 E.g. Koyré, 'La Loi de la Chute des Corps', p. 25.

4 Cf. ch. IV.

PAGE 49

1 Cf. ch. V.

2 'Lettre à Mersenne', 14 août 1634; *Oeuvres*, vol. I, p. 303.

PAGE 50

1 *The Logic of Modern Physics* (New York, 1927), p. 83.

2 *Human Knowledge* (London, 1948), p. 244.

3 Book Review, *Observer*, 4 April 1954.

4 *Op. cit.* (note 2 above), *loc. cit.*

PAGE 51

1 *Religio Laici.* And see Russell's discussion of 'causal ancestry' in *Human Knowledge*, p. 483. Cf. R. B. Braithwaite, *Scientific Explanation* (Cambridge, 1953), pp. 308, 321.

2 'Une intelligence qui pour un instant donné, connaîtrait toutes les forces dont la nature est animée, et la situation respective des êtres qui la composent, si d'ailleurs elle était assez vaste pour soumettre ces données à l'analyse, embrasserait dans la même formule, les mouvements des plus grands corps de l'univers et ceux du plus léger atome: rien ne serait incertain pour elle, et l'avenir comme le passé serait présent à ses yeux. L'esprit humain offre dans la perfection qu'il a su donner à l'astronomie, une faible esquisse de cette intelligence' (Laplace, *Essai philosophique sur les probabilités* (Paris, 2nd ed. 1814), pp. 3–4).

3 This is not to say that research never proceeds in this way, e.g. Kepler, Boyle, Faraday, Röntgen, Mme Curie certainly endured painstaking hunts for 'disturbing factors' in much of their work.

PAGE 52

1 Thus Newton's conclusion of the *Opticks*, Bk. III, I. And cf. Dirac, *Quantum Mechanics* (Oxford, 1930), p. 4.

2 Cf. Galileo, 'Discourses', *Opere*, vol. VII, p. 202 and Newton, 'I have not been able to discover the cause of those properties of gravity...it is enough that gravity does really exist, and act according to the laws which we have explained' (*Philosophiae Naturalis Principia Mathematica* (3rd ed.), Conclusion.

PAGE 53

1 'A chapter of lucky accidents' is Toynbee's happy phrase. Cf. *A Study of History* (Oxford, 1934–54), vol. VII.

2 Alice learned this much from her croquet game: 'It was very provoking to find that the hedgehog had unrolled itself, and was in the act of crawling away' (Lewis Carroll, *Alice in Wonderland* (London, 1897), ch. VIII).

PAGE 54

1 Or is there but one favoured type of causal explanation? What is it? That of classical mechanics? Cf. Du Bois: 'The cognition of nature is the reduction of changes in the material world to motions of atoms, acted on by central forces, independent of time...wherever such a reduction is successfully carried through our need for causality feels satisfied' (*Über die Grenzen des Naturerkennens* (Leipzig, 1872), vol. I). Helmholtz: 'The task of physical science is to reduce all phenomena of nature to forces of attraction and repulsion....Only if this problem is solved are we sure that nature is conceivable' (*Über die Erhaltung der Kraft* (Leipzig, 1847)). Or is it the explanations of quantum mechanics to which all others must be reducible? In his passionate attack on the thesis of this chapter (expressed in my article 'Causal Chains', *Mind*, LXIV, 255) Mr David Braybrooke works hard to miss the point (cf. 'Vincula Vindicata', *Mind*, LXVI, 262). He succeeds completely. Braybrooke writes: 'It would be ridiculous to claim that scientists...do not often confront particular questions to which chain-like causal accounts are appropriate answers' (*ibid.* p. 224). Of course it would. It would also be ridiculous to claim that scientists never confront anything *but* such questions, which was the burden of my article and of this chapter, and which was apparently too much of a burden for Mr Braybrooke to support. As to what scientists do or not do, however, perhaps we ought to hear it from a couple of them, rather than have Mr Braybrooke as our spokesman: 'Consider a wheeled vehicle accelerating on a level road. What is the cause of this motion? For the magistrate it is the driver in charge of the vehicle; for the engineer it is the engine which provides the propulsive power; but for the applied mathematician it is the forward thrust exerted by the road on the wheels or tyres' (Professor G. Temple, 'The Dynamics of the Pneumatic Tyre', in *Endeavour* (1956), p. 200). And, 'In biology...when we speak of

the cause of an event we are really over-simplifying a complex situation. ... The cause of an outbreak of plague may be regarded by the bacteriologist as the microbe he finds in the blood of the victims, by the entomologist as the microbe-carrying fleas that spread the disease, by the epidemiologist as the rats that escaped from the ship and brought the infection into the port' (Professor W. I. B. Beveridge, *The Art of Scientific Investigation* (Heinemann, London, 2nd ed.)).

2 All one *does* do is look and see; that was the argument of ch. I. The operation is psychologically uncomplicated, in one sense, but logically it is complex.

PAGE 55

1 We are not discussing the technical sense of 'wound' as used in surgical theory.

2 Sometimes 'wound' is appropriated for ships and aircraft: 'She plunged on despite her wounds.' But this is pure metaphor.

PAGE 58

1 Just as some school students do not understand the action of Atwood's machine.

PAGE 59

1 This seriously damages Michotte's thesis. Cf. *La perception de la causalité* (Louvain and Paris, 1946).

2 'Of the sense data we cannot know more...than that they are in agreement...' (Leibniz, *Die Philosophische Schriften* (Berlin, 1875–90), vol. IV, p. 356). Contrast 'The elementary proposition consists of names. It is a connexion, a concatenation, of names' (Wittgenstein, *Tractatus*, 4. 22). 'From an elementary proposition no other can be inferred' (*ibid.* 5. 134).

PAGE 61

1 Wittgenstein, *Phil. Inv.* p. 179e.

PAGE 62

1 'The first and oldest words are names of "things"' (Mach, *Science of Mechanics*, p. 579).

PAGE 63

1 No ambiguity results when contexts are thus specified. This parallels the bird-antelope, which is not 'ambiguous' when set into contexts

(ch. I). Cf. Wittgenstein's 'Green is green' (*Tractatus*, 3. 323). 'The silent adjustments to understand colloquial language are enormously complicated' (*Tractatus*, 4. 002).

2 Cf. Wittgenstein, *Tractatus*, 4. 032.

3 *Phil. Inv.* p. 215.

PAGE 64

1 All that matters is that certain systems of concepts can help us to understand what there is. Cf. *Tractatus*, 6. 342 and 6. 3432, two profound observations on the philosophy of physics.

2 Cf. ch. V.

PAGE 65

1 'A hypothesis to be regarded as a natural law must be a general proposition which can be thought to *explain* its instances; if the reason for believing the general proposition is solely direct knowledge of the truth of its instances, it will be felt to be a poor sort of explanation of these instances' (Braithwaite, *Scientific Explanation*, p. 302).

PAGE 66

1 Kepler (*Prodromus Dissertationum Cosmographicarum continens Mysterium Cosmographicum* (in *Johannes Kepler Gesammelte Werke*, München, 1937), I, 10), regards the mathematical harmony underlying observed facts as the cause of the latter. The mathematical harmonies in the mind of the creator furnish the cause 'why the number, the size and the motions of the orbits are as they are and not otherwise'. Galileo too: '[We] cannot understand [physics] if we do not first learn the language and the symbols in which it is written. This book [the universe] is written in the mathematical language... without which one wanders...through a dark labyrinth' (*Opere*, vol. IV, p. 171). (Cf. also *Two Chief World Systems*, pp. 178, 181 ff.) For Descartes the mathematical laws of nature were established by God. Cf. 'Lettre à Mersenne', 15 April 1630 (*Oeuvres* (Cousin), vol. VI, pp. 108 ff.). Boyle as well: 'Mathematical and mechanical principles are the alphabet in which God wrote the world.' 'Nature does play the mechanician' (*Works* (London, 1744), vol. IV, pp. 76 ff.). Cf. also *The Usefulness of Natural Philosophy*, part I, essay 4, 'Containing a requisite Digression concerning those that would exclude the Diety from intermeddling with Matter' (*passim*). Finally Kant in the *Metaphysische Anfangsgründe der Naturwissenschaft* (Leipzig, 1900): 'In every specific natural science there can be found only so much science proper as there is mathematics present in it' (preface).

PAGE 67

1 'The experimenter endeavors to arrange the experiment in such a way that it is most sensitive to one law and as insensitive as possible to all others that play a part, namely by dampening the influence of such circumstances as are governed by the latter' (Weyl, *Philosophy of Mathematics and Natural Science* (Princeton, 1949), p. 153).

2 'Consider, for example, a determination of the mass of an oil-drop in Millikan's experiment to measure the electronic charge: the force is found by Stokes' law from the terminal velocity of fall, the acceleration must be known as the "acceleration of gravity", and this is obtained from the period of oscillation of a pendulum, and so on. Once we understand the *system* of dynamics, we are not compelled to treat the subject like a one-way artery of traffic, as so-called logical presentations of it seem to require' (Watson, *On Understanding Physics* (Cambridge, 1939), p. 117).

PAGE 68

1 'The physicist...accumulates experiences, fits and strings them together by artificial experiments...but we must meet the bold claim that this is Nature with...a good-humoured smile and some measure of doubt...' (Goethe, *Contemplations of Nature*). Cf. Galileo's account of the careful preparation of the inclined-plane experiments, *Dialogues Concerning Two New Sciences* (Dover), pp. 178–9.

PAGE 69

1 'Thus the whole of natural science consists in showing in what state the bodies were when this or that change took place, and that...just that change had to take place which actually occurred' (Euler, *Anleitung zur Naturlehre* (*Opera Omnia*, Lipsiae et Berolini, 1911), vol. VI, § 50).

2 'What we have rather to do is to *accept* the [scientific] language-game, and note false accounts of the matter *as* false' (Wittgenstein, *Phil. Inv.* p. 200e (my insertion)).

PAGE 70

1 Joos, *Theoretical Physics* (Blackie, London, 1951), p. 1.

2 Kolin, *Physics*, 'After establishing a general law by the inductive process, the scientist is enabled to deduce...etc.', p. 21.

PAGE 71

1 Newton, *Principia* (Motte-Cajori), preface, cf. also *Opticks*.

2 Lecture in the series 'The History of Science', given at the University of Cambridge, 4 November 1955.

3 Cf. Braithwaite, *Scientific Explanation*, p. ix.

PAGE 72

1 This is actually what Mrs Beeton is usually misquoted as saying. The original sentiment, however, comes from the writings of Hannah Glasse (fl. 1747) who remarks in *The Art of Cookery* 'Take your hare when it is cased'.

2 '[Kepler] was like the child who having picked a mass of wild flowers tries to arrange them into a posy this way, and then tries another way, exploring the possible combinations and harmonies' (H. Butterfield, *The Origins of Modern Science*, (London, Bell, 1949), p. 56).

PAGE 73

1 Kepler, *Prodromus Dissertationum Cosmographicarum continens Mysterium Cosmographicum*, in *Johannes Kepler Gesammelte Werke* (München, 1937), vol. I.

2 Kepler, *De Motibus Stellae Martis* (in *Johannes Kepler Gesammelte Werke*, München, 1937), vol. III, chs. VI and XXII.

3 *Ibid.* ch. XIII.

4 Kepler, *De Motibus Stellae Martis*.

5 *Ibid.* chs. XXII, XXIII.

PAGE 74

1 *Ibid.* pp. 177, 178.

2 Aristotle (Oxford, trans. Ross, etc.): *De Caelo*, 268b, 276b, 'Circular motion is necessarily primary...the circle is a perfect thing...the heavens complete their circular orbit...the heaven...must necessarily be spherical'. Also 286b, 289b. *De Generatione et Corruptione*, 337a, 'Circular motion...is the only motion which is continuous'. 338a, 'It is in circular movement...that the "absolutely" necessary is to be found' (*Physica*, 223b, 227b, 265b, 248a, 261b, 262a). 'The circle is the first, the most simple, and the most perfect figure' (Proklus *Commentary on Euclid's Elements* (London, T. Payne, 1788–9), definitions XV and XVI). Cf. also Dante 'Lo cerchio é perfettissima figura' (*Convivio* (Torino, 1927), II, 13).

Cf. Galileo, *Two Chief World Systems*...(1632) (California, 1953), pp. 10–60, especially: 'If such a motion [rectilinear] belonged by nature to a body, then from the beginning it would not be in its natural place; hence the ordering of the world's parts would not be a perfect one.

We assume however, that the ordering of the world is perfect; consequently, it cannot by nature be intended to change place, nor consequently, can it be intended to move in a straight line.' Cf. also Hobbes, *De Corpore Politico* (London, 1652), part III, and M. H. Nicolson, *The Breaking of the Circle* (Evanston, 1950).

PAGE 75

1 Ch. XLIV.

2 'Nec aequationes Physica computatae, observatis (quas vicaria hypothesis repraesentat) consentiebant' (p. 285).

3 'Quid ergo dicendum?...Orbitae Planetae non est circulus, sed ingrediens ad latera utraque paulatim, iterumque ad circuli amplitudinem in perigaeo exiens. Cujusmodi figuram itiniris ovalem appellitant' (p. 286).

4 'Atque ex hoc quoque demonstratum, quod supra cap. XX, XXIII promisi me facturum: Orbitam Planetae non esse circulum sed figurae ovalis' (p. 287).

5 *Collected Papers* (Harvard, 1931), vol. I, 73.

PAGE 76

1 Ch. XLV.

2 'Cogita ipse lector, et vim argumenti persentices. Quia non putavi fieri ullo alio medio posse, ut Planetae orbita redderetur ovalis. Haec itaque cum ita mihi incidissent, plane securus de quantitate hujus ingressus ad latera, nimirum de consensu numerorum, jam alterum de Marte triumphum egi.' And even 'Ac nos, bone lector, par est triumpho tam splendido dieculum unam...indulgere, cohibitis interea novae rebellionis rumoribus, ne apparatus iste nobis citra voluptatem pereat' (p. 290).

3 My friend and colleague, Dr A. R. Hall, has appreciated the significance of Kepler's first non-circular orbit hypothesis, the oviform curve. Cf. *The Scientific Revolution* (London, 1955), p. 125. But Hall does not mark the importance of the ellipse as a *mathematical tool*, even in this early phase of Kepler's Martian work. It is an object of this chapter to distinguish two strands in Kepler's thought at this critical stage: the *physical* hypothesis of the oviform orbit, and the *mathematical* hypothesis of the elliptical curve.

4 'Quocunque dictorum modorum delineetur linea corpus Planetae possidens, sequitur jam, viam hanc, punctis $\delta, \mu, \gamma, \sigma, \pi, \rho, \lambda$ signatum, *vere esse ovalem, non ellipticam*, cui Mechanici nomen ab ovo ex abusu collocant. Ovum enim duobus turbinatum verticibus, altero tamen obtusiori, altero acutiori, et lateribus inclinatus cernitur. Talem figuram dico nos creasse' (p. 295, my italics).

PAGE 77

1 'Tot caussis concurrentibus apparet resegmentum nostri circuli eccentrici infra multo esse latius, quam supra, in aequali ab apsidibus recessu. Quod cuilibet vel numeris exploratu facile est, vel Mechanica delineatione, assumpta evidenti aliqua eccentricitate' (p. 296).

2 Cf. Small, *The Astronomical Discoveries of Johannes Kepler* (London, 1804): '[Kepler] had considered the oval as a real ellipse' (p. 303). Mill: 'Kepler swept all these circles away, and substituted the conception of an exact ellipse' (*A System of Logic* (8th ed.), p. 195). Peirce: 'The question is whether [the planet moves as it ought to do] ...owing to an error in the law of areas or to a compression of the orbit. He ingeniously proves that the latter is the case' (*Collected Papers*, p. 73). Wolf also obscures the important issue. He writes: 'Only after trying many ovals, all larger at one end than at the other, did it occur to Kepler to try an *ellipse*, the simplest form of oval. He eventually arrived, by trial and error, at an elliptic orbit' (*A History of Science, Technology and Philosophy* (London, 1952, 2nd ed.), p. 139). For one thing, it is questionable whether an ellipse is best described as 'the simplest form of oval'. For another, as this chapter purports to make clear, to say only that the elliptical orbit was arrived at by trial and error is to obscure the important function of the ellipse even *while* the physical hypothesis of the oviform curve is dominating Kepler's thinking. Even the great Dreyer is not as lucid at this juncture as one might have wished: 'For finding the areas of the oval sectors Kepler substituted for the oval an ellipse' (*History of the Planetary Systems from Thales to Kepler* (Cambridge, 1906), p. 390). But the considerations which allowed this substitution—which made it seem plausible to Kepler—are not discussed.

3 P. 296. The physical significance of the *plani oviformis* for Kepler is clear: after noting that Mars' calculated positions fall within the perfect circle, he straightway treats the resulting oviform as the joint-product of two physical attractive forces; that of Mars and of the sun. At this stage he never treats the ellipse in this way.

PAGE 79

1 This diagram (*op. cit.* p. 291) is the first non-circular curve to appear in *De Motibus Stellae Martis*, though it is hinted at on p. 276. The curve $\delta D\lambda$ is clearly half of a perfect ellipse. But Kepler is not thinking of it here as a possible Martian orbit. The latter, as he stresses five pages later, 'esse ovalem, non ellipticam' (p. 295). This is the geometrical curve to which the oviform approximates; such a device allows Kepler to enlist the help of Archimedes (p. 297). No geometer could do much with the oviform *per se*: 'Ovum enim duobus turbinatum

verticibus, altero tamen obtusiori, altero acutiori, et lateribus inclinatus cernitur.'

2 'Si figura nostra esset perfecta ellipsis', and 'Sit autem haec figura perfecta ellipsis, parum enim differt. Videamus quid inde sequatur' (p. 297).

3 Kepler actually puts the same point the other way round: 'Concessis itaque, quae posuimus, quod planum ellipsis a plano nostri ooidis insensibiliter differat, eo quod compensatio sit inter supernos excessus ooidis supra ellipsin, et infernos defectus' (*De Motibus*, p. 299).

In different ways Small and Hall miss the significance of these passages. Small identifies the oval with an ellipse: '[Kepler] had considered the oval as a real ellipse' (*op. cit.* p. 303). Hall distinguishes the oval and the ellipse so thoroughly that they seem to be unrelated in this phase of Kepler's work: 'It was the accidental observation of a numerical incongruity that led [Kepler] to substitute for the oval an ellipse' (*The Scientific Revolution*, p. 125). Here, in a spectacular triumph for the history of physics, a complex, intractable phenomenon is made a subject for thought by regarding it as an approximation to a simple, easily managed mathematical entity. We must distinguish the supposed physical orbit and the mathematical tool, as Small does not. But we must not divorce them as Hall is in danger of doing.

4 P. 297.

5 P. 299. In an obscure note Dreyer says: 'The sun is not in one of the foci of this auxiliary ellipse' (*Planetary Systems*, p. 390). But he is just wrong, since the quotations just given, along with the previous diagram, lack sense unless the sun is taken to reside in one of the foci of this 'auxiliary' ellipse.

PAGE 80

1 P. 367, ch. LIX, which argues that the librations of Mars in its supposed epicycle really gives a perfect elliptical orbit.

PAGE 81

1 'Argumentatio mea talis fuit, qualis cap. XLIV, L, et LVI. Circulus cap. XLIII peccat excessu, ellipsis capitus XLV peccat defectu. Et sunt excessus ille et hic defectus aequales. Inter circulum vero et ellipsin nihil mediat nisi ellipsis alia. Ergo ellipsis est Planetae iter.' P. 366. Dreyer is the only historian I know of who has seen the real meaning of this passage. He writes: 'The true orbit was therefore clearly proved to be situated between the circle and the oval' (*Planetary Systems*, p. 391).

2 Despite Kepler's misleading reference, Dreyer rightly distinguishes the 'oval' and the 'auxiliary' ellipse—something which Small (cf. n. 2,

p. 197) failed to do and which Hall (cf. n. 3, p. 198) completely overdoes.

PAGE 82

1 Ch. LVI.

2 P. 345.

3 Cf. p. 346 and diagram.

4 Ch. LVIII, p. 364.

5 P. 366. As Dreyer says: 'This compelled Kepler to return to the ellipse, which he had already employed as a substitute for the oval' (*Planetary Systems*, p. 392).

6 'Patet igitur, viam buccosam esse; non igitur ellipsin. Ac cum ellipsis praebeat justas aequationes, hanc igitur buccosam, jure injustas praebere.'

PAGE 83

1 'Itaque ne hic quidem valde haesi. Multo vero maximus erat scrupulus, quod pene usque ad insaniam considerans et circumspiciens, invenire non poteram, cur Planeta, cui tanta cum probabilitate, tanto consensu observatarum distantiarum, libratio *LE* [cf. diagram on p. 365] in diametro *LK* tribuebatur, potius ire vellet ellipticam viam, aequationibus indicibus. O me ridiculum! perinde quasi libratio in diametro, non possit esse via ad ellipsin. Itaque non parvo mihi constitit ista notitia, juxta librationem consistere ellipsin; ut sequenti capite patescet: ubi simul etiam demonstrabitur, nullam Planetae relinqui figuram Orbitae, praeterquam perfecte ellipticam; conspirantibus rationibus, a principiis Physicis, derivatus, cum experientia observationum et hypotheseos vicariae hoc capite allegata' (*ibid.*).

2 Pp. 367–424. Here is where the hypothetico-deductive account of a physical theory has a point. Kepler has solved his physical problem; he has caught his *explicans*. Now he must elaborate it deductively. He does in fact do this, but any of his competent students could have done it for him, the pattern was now clear. But no one but Kepler could have written the first 367 pages of *De Motibus Stellae Martis*.

PAGE 84

1 Cf. next chapter.

2 Mill, *A System of Logic*, Bk. III, ch. II, §3.

3 Cf. Whewell, *Philosophy of Discovery*, and *Novum Organum Renovatum* (London, 1858).

4 Braithwaite, *Scientific Explanation*, p. ix.

PAGE 85

1 Peirce, *Collected Papers*, vol. I, p. 31.

2 Preliminary to the discussion following must be an appreciation of the logical distinction between (1) reasons for accepting an hypothesis H, and (2) reasons for suggesting H in the first place. (1) is pertinent to what makes us say H is true, (2) is pertinent to what makes us say H is plausible. Both are the province of logical inquiry, although H-D theorists discuss only (1) saying that (2) is a matter for psychology or sociology—not logic. This is just an error. What leads to the initial formation of H—the 'click', intuition, hunch, insight, perception, etc.—this *is* a matter of psychology. But many hypotheses flash through the investigator's mind only to be rejected on sight. Some are proposed for serious consideration, however, and with good reasons. Kepler would have had good *reasons* for rejecting the hypothesis that Jupiter's moons cause the apparent accelerations of Mars at 90° and 270°. He also had good reasons for proposing that *all* the planets move in ellipses (after having established only that Mars does). This analogical type of hypothesis though, could not possibly *establish* that all planets move in ellipses. We are discussing the rationale behind the proposal of hypotheses as possible *explicantia.* H-D theorists never raise the problem at all.

3 By Jenkinson; cf. *Prior Analytics* (Oxford, ed. Ross), vol. II, p. 25.

4 *Op. cit.* vol. II, p. 25. Cf. also *Posterior Analytics*, vol. II, p. 19. Cf. 'The particular facts are not merely brought together but there is a new element added to the combination by the very act of thought by which they are combined.... The pearls are there, but they will not hang together till someone provides the string' (Whewell, *Novum Organum Renovatum*, pp. 72, 73).

5 Vol. V, § 145.

6 *Op. cit.* § 146.

7 § 171.

PAGE 86

1 §§ 173, 174.

2 § 194.

3 § 188.

4 In 1867, cf. *Collected Papers*, vol. II, bk. III, ch. 2, part III.

PAGE 87

1 Cf. Braithwaite, *Scientific Explanation*, p. ix, ll. 7–11. Cf. Peirce: 'How was it that the man was ever led to entertain that true theory?

You cannot say that it happened by chance' (*op. cit.* vol. v, § 591), and 'How few were the guesses that men of surpassing genius had to make before they rightly guessed the laws of nature' (vol. v, § 604).

2 Cf. Goethe: 'In science all depends on what is called an *aperçu*, on a recognition of what is at the bottom of the phenomena' (*Geschichte der Farbenlehre* (*Werke*, Weimar, 1887–1918) (4, 'Galileo')), and Einstein: 'The discovery of these elemental laws...is helped by a feeling for the order lying behind the appearance' (preface to Planck's *Where is Science Going*? (London, 1933)).

PAGE 88

1 Cf. ch. VI.

PAGE 89

1 Peirce, *op. cit.* §§ 581 and 602.

2 The classical studies on this subject are those of Duhem: *De l'accélération produite par une force constante*, Congrès International de Philosophie (Genève, 1905); *Études sur Léonard de Vinci; Les précurseurs Parisiens de Galilée* (Paris, 1913).

PAGE 90

1 'When we wish to introduce ideas whose connexion is represented in a mathematical law, we cannot first introduce the ideas and then impose the law on the symbols representing the magnitudes involved, for until we have the law the ideas are not made clear and definite. The numbers of arithmetic are not *entities* on which the laws of arithmetic are imposed....The description of any dynamical phenomenon is always relative to some system of reference' (Watson, *On Understanding Physics*, p. 120).

2 'The experimental verifications are not the basis of the theory, but its culmination' ('Physique et metaphysique', *Revue des questions scientifiques*, XXXVI (1897)).

PAGE 91

1 This distinction is already clear as early as Eudoxos (cf. Aristotle, *Metaphysica*, Λ 1073b17—1074a14; cf. also Cohen and Drabkin, *Source Book in Greek Science* (New York, 1948), p. 102, ll. 18–19). Aristotle also makes the point independently (*Physica*, II, 1, 193b, 22), being as dissatisfied with the abstract calculating devices of Eudoxos and Kallipus as Huygens and Leibniz were with Newton's *Principia*. Geminus is more forceful still (*Elementa astronomiae*, ed. Manitius (Leipzig, 1898), p. 283). Apollonius, Hipparchus, and Claudius Ptolemy, while well aware of this distinction, decide in favour

of inventing abstract mathematical devices. They view the provision of an *explanation* of celestial motions as beyond human powers (*Almagest IX*, 2: XIII, 2: III: VI, 13: IX, 3, 4: XI, 10: XII, 8—also in Halma edition pp. 41–2; and see *Hypotyposes* (Paris, Halma, 1800), p. 151). Aquinas would have regarded the *Principia* as he did the astronomy of his own day, as the 'Assuming of an assertion tentatively [for the purpose of] deriving results from it which can be compared with our observations' (*Summa Theologica* (*Opera omnia*, Rome, 1882–1948), t. 4–12).

So too Osiander would have treated Newton's theory as 'hypotheses [that are] but bases of calculation. Even if they are false it does not matter much provided that they describe the observed phenomena correctly' (*Letter to Copernicus* (1541)).

2 Newton, *Opera* (J. Nichols, London, 1779), vol. IV, pp. 310, 318ff., 320, 324ff., 328, 335. Cf. Clarke who speaks of 'The false philosophy of the materialists who oppose the *mathematical* principles of philosophy' (*Clarke-Leibniz Letters* (Knapton, London, 1717), first reply).

This was precisely the difference between Galileo's and Descartes' treatment of free fall in ch. II. The former sought a formal principle which would mathematically pattern the observed data. The latter sought the cause which was physically responsible for the data being as observed.

3 Helmholtz, *Über die Erhaltung der Kraft* (Leipzig, 1847). Cf. ch. III, n. 1 (second entry), p. 191.

4 Broad, *Perception, Physics and Reality* (Cambridge, 1914), p. 276.

5 Maxwell, *Scientific Papers* (London, 1890, reprinted Dover). In several papers models of the aether are suggested; one is made of coil springs and leather strips. Cf. Kelvin: 'I never satisfy myself until I can make a mechanical model of a thing. If I can make a mechanical model I can understand it. As long as I cannot make a mechanical model all the way through, I cannot understand it.... I want to understand light as well as I can without introducing things that we can understand even less of' (quoted in Bridgman, *Logic of Modern Physics*, p. 45).

PAGE 92

1 'One may, however, extend the meaning of the word "picture" to include *any way of looking at the fundamental laws which makes their self-consistency obvious*' (Dirac, *Quantum Mechanics*, p. 10).

PAGE 93

1 'So auch sagt es nichts über die Welt aus, dass sie sich durch die Newtonische Mechanik beschreiben lässt; wohl aber, dass sie sich so

durch jene beschreiben lässt, wie dies eben der Fall ist' (Wittgenstein, *Tractatus*, 6. 342).

2 In the physics of molar bodies moving at moderate speeds through 'middle sized' spaces.

3 Cf. e.g. *The Foundations of Science*, p. 28 (l. 5), pp. 97, 99, 102, 106, 125, 318, 328.

PAGE 94

1 'Corpus omne perseverare in statu suo quiescendi vel movendi uniformiter in directum, nisi quotenus a viribus impressis cogitur statum illum mutare' (Newton, *Philosophiae Naturalis Principia Mathematica*, 'Axioms').

2 Aristotle, *De Caelo* (trans. Ross, Oxford University Press, 1928), 276a (22ff.), 277a (14ff.), 294b (32ff.); *Physica*, 256a (5–21),esp. 11, 29 and 30, 256b, 258b (10ff.) 260a (l. 12ff.), esp. 265a (13), and ll. 28–35, 266b (25–35) and 267a.

PAGE 95

1 Newton, *Principia*. Cf. Cote's preface to the 2nd ed. (1713).

2 Cf. Galileo, *Dialogues Concerning the Two Chief World Systems* (Univ. of California, 1953). The First Day, esp. pp. 45, 52; the Second Day, esp. pp. 113, 115ff., 188ff., 248, 253, 257ff.; the Third Day, esp. pp. 320ff., etc.

3 Atwood and Whewell and Lagrange may be cited; the Royal Society of the late eighteenth and early nineteenth centuries was full of physicists of this temperament.

4 Cf. esp. Helmholtz: 'The task of physical science is to reduce all phenomena of nature to forces of attraction and repulsion the intensity of which is dependent only upon the mutual distance of material bodies. Only if this problem is solved are we sure that nature is conceivable' (*Über die Erhaltung der Kraft*). Cf. Euler: 'The whole of Natural Science consists in showing in what state the bodies were when this or that change took place, and that,...just that change had to take place which actually occurred' (*Anleitung zur Naturlehre* (in *Opera Posthuma*, II, Leipzig and Berlin, 1911), vol. VI, § 50).

PAGE 96

1 Cf. Galileo, *Two Chief World Systems*, pp. 28–32.

2 This type of reasoning led to the discovery of Neptune by V. J. Leverrier (1846), the greatest triumph of Newton's mechanics. The companion stars of Sirius and Procyon were also discovered before they were seen. (Incidentally, the philosophical examination of Leverrier's work has yet to be written. How remarkable that this man should

have raised classical mechanics to its highest pinnacle by predicting the unseen Neptune as being responsible for observed aberrations in the orbit of Uranus; yet by this same argument he postulated the 'planet' Vulcan to explain Mercury's precessions at perihelion, and classical mechanics met its most telling failure. It cannot have been the fate of many physicists to have served both as the saviour and as the executioner of a physical theory.)

Cf. Hertz (*Principles of Mechanics*, Bk. II, p. 735): 'At first it might have appeared that the fundamental law was far from sufficient to embrace the whole extent of facts which nature offers us.... We saw, however, that we could also investigate abnormal and discontinuous systems if we regarded their abnormalities and discontinuities as only apparent; that we could also follow the motion of unfree systems if we conceived them as portions of free systems; that, finally, even systems apparently contradicting the fundamental law could be rendered conformable to it by admitting the possibility of concealed masses in them.'

3 Cf. A. Pap, *The A Priori in Physical Theory* (New York, 1946), pp. 1–55; V. F. Lenzen, *Physical Theory* (New York, 1931), pp. 10–15 and parts I and II.

PAGE 97

1 When, that is, the substance is free of impurities and when the experiment is free of abnormalities. These conditions will be assumed.

2 Thus Hertz writes: 'We consider the problem of mechanics to be to *deduce* from the properties of a material system which are independent of the time, those phenomena which take place in time and the properties which depend on the time. For the solution of this problem we lay down the following, and only the following, fundamental law, *inferred from experience*' (*Principles of Mechanics*, vol. II, p. 308, my italics).

PAGE 98

1 Cf. James: 'All the magnificient achievements of mathematical and physical science...proceed from our indomitable desire to cast the world into a more rational shape in our minds than the shape into which it is thrown there by the crude order of our experience' (*Essays in Pragmatism* (Hafner, 1948), p. 38).

Sigwart: 'That there is more order in the world than appears at first sight is not discovered *till the order is looked for*' (*Handbuch zu Vorlesungen über die Logik* (Tübingen, 1835), Bd. II, 5. 382).

The readings which supported Boyle's law would have supported a number of other correlations as well. We regard the readings as describing two intersecting curves, as in (*a*) below.

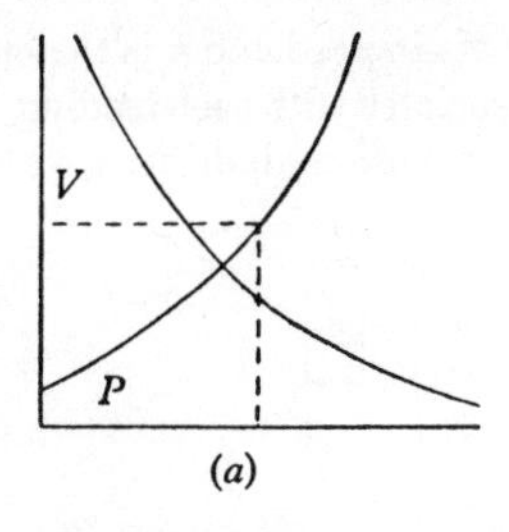

(*a*)

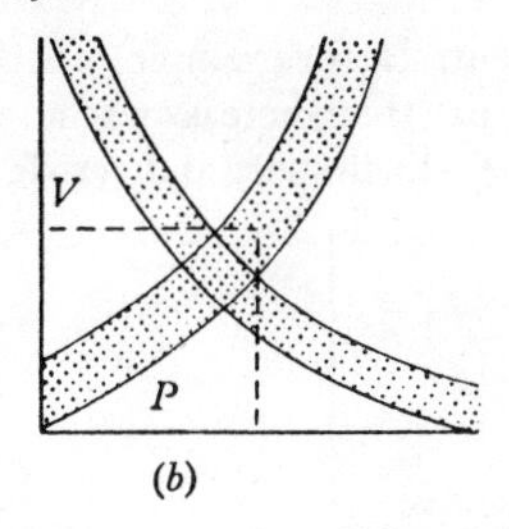

(*b*)

This representation is possible only after the readings have been thoroughly corrected. If anything, the readings describe two 'belts' or 'ribbons' of possible data, as in (*b*) above.

How could a laboratory measurement approximate to a number, which is logically sharp? Contrast Kepler's geometrically clean ellipse with Tycho's 'fuzzy' data. You can only get within the neighbourhood of a number or of an ellipse. Volume lies between two points, covers an interval. In 'verifying' Boyle's Law one always gets 'areas' of readings, never a line. And of an infinite number of curves that can pass through that area the physicist chooses one. [Even here Boyle never made the choice. His successors drew the curve through his data. As Leibniz remarks: 'What a pity that, for all his hundreds of experiments, [Boyle] provides no new general ideas for the interpretation of nature' (*Die Philosophischen Schriften*).]

Cf. a paper by A. H. Yates in *Flight* (14 June 1957), where the above considerations are generalized:

'If values of some quantity, y, are measured for a range of values of another variable, x, we are faced with the problem of drawing a curve through a series of points, as in Fig. 1:

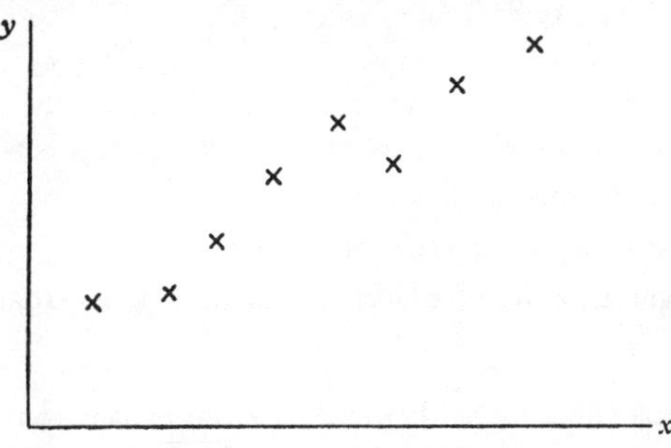

Fig. 1

What kind of curve should be drawn?...Doubts often arise because the observer fails to realize that each cross is merely the centre of an area. He should constantly have in mind the accuracy of his measurement. If, for example, y is the reading of airspeed and x is the engine thrust, then there is an accuracy to be associated with each reading, such as $\pm$ 1 kt. in speed and $\pm$ 10 lb. in thrust. If these are indicated, we have:

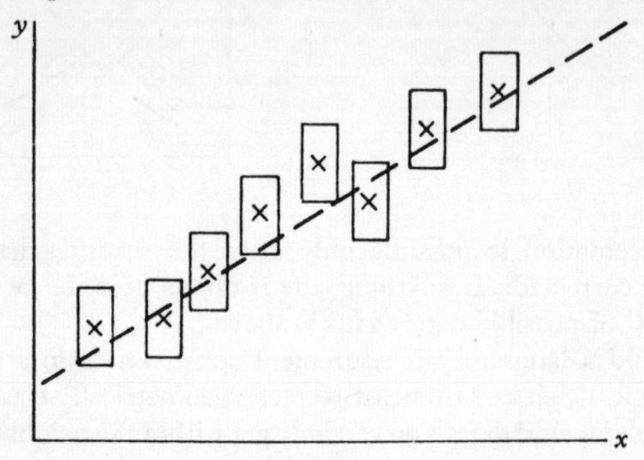

Fig. 2

and there is little to be said for drawing anything more elaborate than a straight line' (p. 809).

Cf. also 'Accuracy and Commonsense' by the present author, in *Flight* (28 June 1957), p. 856.

PAGE 99

1 'Mutationem motus proportionalem esse vi motrici impressae, et fieri secundum lineam rectam qua via illa imprimitur' (Newton, *Principia*, 'Axioms').

2 Cf. e.g. Broad, *Scientific Thought* (London, 1923), p. 162. Cf. Mach, *Science of Mechanics* (Open Court, 1942), pp. 244–6, and Joos, *Theoretical Physics*, p. 82, l. 4.

3 m is assumed constant. A more cautious formulation would be $F = d/dt(mv) = du/dt$, which leaves open the question of the constancy of mass (cf. W. Kaufmann in *Göttinger Nachrichten*, 8 Nov. 1901).

4 Newton, *Principia*, p. xvii, ll. 26–7.

5 *Ibid.* p. 5 (last para.) and p. 192, Scholium.

6 As Broad suggests, *Perception, Physics and Reality*, p. 349, ll. 8–10.

PAGE 100

1 These accounts (1–5) are neither exclusive nor exhaustive.

PAGE 101

1 By the second law, the unbalanced force on m_1 is $F - m_1 g = m_1 a$. The unbalanced force on m_2 comprises its weight $m_2 g$, plus the upward pull (F) of the string. This can be expressed as $m_2 g - F = m_2 a$.

Thus
$$F = m_2 g - m_2 a$$
$$= m_2(g - a).$$
Substituting this in
$$F - m_1 g = m_1 a;$$
$$m_2(g - a) - m_1 g = m_1 a,$$
or
$$m_2 g - m_2 a - m_1 g = m_1 a.$$
This is the same as
$$a(m_2 + m_1) = g(m_2 - m_1),$$
and from this it follows that
$$a = g\frac{m_2 - m_1}{m_2 + m_1}.$$

2 A. Kolin, *Physics* (McGraw-Hill, 1950), pp. 46–7. Cf. Humphreys and Beringer, *Atomic Physics* (Harper, New York, 1950), 'Newton's laws are not physical laws...but are definitions of the basic concepts in dynamics...Newton's second law provides us with a working definition of force' (pp. 38–9). 'These laws are the foundation of ordinary dynamics and simultaneously form the basic definitions, hypotheses and deductions of the theory' (*ibid.* p. 37). Clerk Maxwell, *Matter and Motion* (London, 1920), '"Impressed" force...is completely defined and described in Newton's three laws of motion...' (III, 40, p. 27). Poincaré, *Science and Hypothesis* (London, 1905), 'This [second] law of motion...ceases to be regarded as an experimental law, it is now only a definition' (p. 100); and 'Les principes de la dynamique nous apparaissent d'abord comme des vérités expérimentales; mais nous avons été obligés de nous en servir comme définitions', 'Des Fondements de la Géometrie', in *Revue de Meta-Physique et de la Morale* (1899), p. 267.

3 Cf. Newton, *Principia*, '[This] will make the system of the two bodies ...to go forwards *in infinitum* with a motion continually accelerated; which is absurd and contrary to the first law' (p. 25). (The first law is, of course, only a special case of the second, i.e. where $\Sigma F = 0$.) Similar arguments can be found throughout the *Principia*.

4 G. Atwood, *A Treatise on the Rectilinear Motion and Rotation of Bodies, with a Description of Original Experiments Relative to the Subject* (Cambridge, 1784). Cf. p. 4, 'The laws of motion have been esteemed not only physically but mathematically true'.

5 *Ibid.* p. 30.

6 *Ibid.* p. 33, 'Many experiments have been produced...to disprove the Newtonian measure...it immediately belongs to the present subject to determine whether the conclusions which have been drawn

from these experiments arise from any inconsistency between the Newtonian measures of force and matter of fact' (p. 30).

7 In this he only obeyed Newton's dicta: 'The qualities of bodies are only known to us by experiments, we are to hold for universal all such as universally agree with experiments' (Bk. III, Appendix, Rule Three). 'We are to look upon propositions inferred by general induction from phenomena as accurately or very nearly true' (Rule Four). 'Particular propositions are inferred from the phenomena, and afterwards rendered general by induction' (p. 547). 'Analysis consists in making experiments and observations and in drawing general conclusions from them by induction' (*Opticks* (1721), p. 380).

8 Cf. *Treatise*, pp. 298ff.

PAGE 102

1 Broad, *Scientific Thought*, p. 162, and cf. pp. 163 (bottom), and 164 (ll. 16–18).

2 '[In] the laws of Nature...there appears...not the least shadow of necessity....These therefore we must...learn...from observations and experiments....All sound and true philosophy is founded on the appearances of things' (*Principia*, Cote's preface). 'Such principles ...are confirmed by abundance of experiments' (p. 21, Scholium). Cf. further references in the *Principia*, p. 24, ll. 28–9, pp. 325–6, p. 398, p. 294. '*The laws observed during the motion of bodies acted on by constant forces* admit of easy illustrations from matter of fact' (my italics). Also p. 308, ll. 21–3 and p. 329, para. III.

3 *Treatise*, p. 2.

4 *Treatise*, p. 279.

PAGE 103

1 W. Whewell, *Astronomy and General Physics* (London, 1834), pp. 211, 212 (my italics).

2 Cf. Wittgenstein, *Tractatus*, 6. 342.

PAGE 104

1 The synthetic *a priori* view of the laws of mechanics, first articulated by Kant (*Kritik der Reinen Vernunft*, Leipzig, 1781), and later by Natorp (*Die Logische Grundlagen der Exakten Wissenschaften* (Berlin, 1910)), and Cassirer (*Determinismus und Indeterminismus in der Modernen Physik*, Goteborg, 1937), is not just a quaint philosopher's invention. It was an important, if misguided, attempt to do justice to actual uses of laws in physics. Physicists still make the attempt: thus Peierls writes: 'People sometimes argue whether Newton's second law

is a definition of force or of mass, or whether it is a statement of an objective fact. *It is really a mixture of all these things'* (*The Laws of Nature* (Allen and Unwin, 1955), p. 21). Cf. Broad: 'This mixture of convention and observation is a very common feature in scientific laws' (*Scientific Thought*, p. 160). But dynamical laws are not *mixtures*. '$F=m(d^2r/dt^2)$' can be used in physics in different ways. This does not make the second law a *mixture*, whatever that may mean. Cf. Weyl: 'In its place [i.e. the clear-cut division into *a priori* and *a posteriori*] we have a rich scale of gradations of stability' (*Philosophy of Mathematics and Natural Science*, p. 154).

2 The *Principia* abounds with this use of the second law: Cf. e.g. p. 14 (Cor. I, l. 9), p. 17 (Cor. III, l. 2), p. 20 (Cor. V, l. 5), p. 21 (Cor. VI, l. 2 and Schol. l. 3), p. 25 (ll. 31–2), p. 42 (Prop. II, ll. 1 and 6), p. 44 (ll. 1, 6, 14), p. 136 (l. 9), p. 162 (l. 6), p. 164 (l. 6), p. 166 (l. 19), p. 169 (ll. 2, 11), etc., and see esp. p. 244 (Schol. l. 8), p. 327 (Section VII, l. 11), p. 368 (l. 1), p. 410 (l. 13), p. 414 (last l.), and p. 442 (l. 11).

3 *Perception, Physics and Reality*, p. 322, ll. 20–5.

4 In *Scientific Thought*, p. 165.

5 *Principia*, Section VI, pp. 303–26.

6 Wittgenstein seems at times to suggest this: 'Newtonian Mechanics ... brings the description of the universe to a unified form' (*Tractatus*, 6. 341). 'Mechanics determines a form of description by saying: All propositions in the description of the world must be obtained in a given way from a number of given propositions—the mechanical axioms.'

The position is adopted explicitly by Watson: 'What we have called the laws of nature are the laws of our methods of representing it' (*On Understanding Physics* (Cambridge Univ. Press, 1939), p. 52); and by Toulmin: 'Laws of Nature do not function as premises *from which* deductions to observational matters are made, but as rules of inferences *in accordance with which* empirical conclusions may be drawn from empirical premises' (*The Philosophy of Science* (Hutchinson's University Library, 1953)).

7 *Scientific Thought*, p. 165.

PAGE 105

1 *Perception, Physics and Reality*, pp. 336–7.

PAGE 107

1 Kepler's laws would be fulfilled precisely when M tends to infinity, γ tends to zero, and γM tends to a finite limit.

2 Consider Kant's first published work, *Thoughts on the True Estimation of Living Forces* (1746–9), in which he writes: 'This law [of gravitation] is arbitrary...God could have chosen another, for instance the inverse threefold relation' (para. 10). And cf. Whewell: 'No reason, at all satisfactory, can be given why such a law [of gravitation] should of necessity be what it is; but...very strong reasons can be pointed out why, for the beauty and advantage of the system, the present one is better than others' (*Astronomy and General Physics*, p. 216).

3 The moon's distance is sixty times the earth's radius. If the law of gravitation is true, the moon is deflected from its inertial motion $1/60^2$ of the distance a body falls in a second at the earth's surface. This distance is 193 in. $1/60^2$ of this is 0·0535 in., very close to the recently observed 0·0534 in.

PAGE 109

1 Cf. Russell's remark: 'The reason for accepting an axiom...is...that many propositions which are nearly indubitable can be deduced from it, and that no equally plausible way is known by which these propositions could be true if the axiom were false, and nothing which is probably false can be deduced from it' (*Principia Mathematica* (Cambridge, 1925), 2nd ed., vol. 1, p. 59, VII).

PAGE 110

1 Cf. e.g. Hertz, *Die Principien der Mechanik in Neuem Zusammenhange Dargestellt* (Leipzig, 1894). Mach writes: 'In the beautiful ideal form which Hertz has given to mechanics, its physical contents have shrunk to an apparently almost imperceptible residue. It is scarcely to be doubted that Descartes...would have seen in Hertz' mechanics...his own ideal' (*Mechanics*, p. 323, cf. also p. 324).

2 Whewell, *Astronomy and General Physics*, pp. 226 ff.

3 The apsidal line ought, according to the *Principia*, move round once in eighteen years. It is observed to do so twice in that period of time.

PAGE 111

1 *Phil. Trans. Roy. Soc.* (1775).

2 *Phil. Trans. Roy. Soc.* (1798), pp. 469–526.

3 *Ibid.* p. 484.

4 *Ibid.* p. 522.

5 'On the Newtonian Constant of Gravitation' (*Phil. Trans. Roy. Soc.* (1895), pp. 1–72). Boys' work culminated a series of experiments aimed at the determination of the attractive force between any two point masses, beginning with Michell and later Cavendish (*Phil. Trans. Roy. Soc.* (1798)), through Reich (*Comptes Rendus* (1837)),

Baily (*Phil. Mag.* (1842)), Cornu and Baille (*Comptes Rendus*, vols. LXXVI and LXXXVI). All of these physicists treat $F=\gamma(Mm)/r^2$ as a straightforward empirical proposition.

6 *Ibid.* p. 70.

PAGE 116

1 This is the law of the impossibility of perpetual motion of the first type (mechanical). It is itself a restricted version of the law of conservation of energy [$T_2+U_2=T_1+U_1=constant$—where kinetic energy $\frac{1}{2}mv^2$ is denoted by 'T', and potential energy by 'U']. This in turn is only a first integral of Newton's second law, since only $\dot{v}$ (i.e. ds/dt) appears.

Cf. 'If the devisers of new machines, who made such futile attempts to construct a *perpetuum mobile*, would acquaint themselves with this principle, they would... understand that the thing is utterly impossible by mechanical means' (Huygens, *Horologium Oscillatorium, Sive De Motu Pendulorum ad Horologia Aptato Demonstrationes Geometricae* (Paris, 1673)).

2 This is another formulation of the second law of thermodynamics.

The highest attainable efficiency for an ideal heat engine (a Carnot engine) requires an expenditure of work as follows:

$$W=\eta Q_1=\eta Q_3/(1-\eta).$$

A super-Carnot engine with an efficiency higher than η would, when coupled with a Carnot engine, give us the practical equivalent of a *perpetuum mobile* of the second kind. But experience and thermodynamic theory (supported now by quantum theory) show that such an engine is impossible.

3 $$U=C\left[1\frac{(1-v/c)(1-w/c)}{1+vw/c^2}\right].$$

PAGE 119

1 *The Method of Theoretical Physics* (Oxford, 1933); and cf. Goethe, *Geschichte der Farbenlehre*, § 4.

PAGE 120

1 Even this may help. Were one undecided between an atomic and an opposing theory, e.g. Descartes', it might help to learn of these unit-constituents of chlorine.

Cf. 'Ac proinde si quaeratur quid fiet, si Deus sufferat omne corpus quod in aliquo vase continetur, et nullum aliud in ablati locum venire permittat? respondendum est, vasis latera sibi invicem hoc ipso fore contigua. Cum enim inter duo corpora nihil interjacet, necesse est ut se mutuo tangant, se manifeste repugnat ut distent, sive ut inter ipsa sit distantia, et tamen ut ista distantia sit nihil; quia omnis distantia est modus extensionis, et ideo sine substantia extensa esse non potest' (Descartes, *Principia* (vol. II, p. 18)). Cf. also *Les Météores*, in *Oeuvres*, vol. VI, p. 238; Mouy, *La physique cartésienne* (Paris, 1934), pp. 101–6.

2 'The parts of all homogeneal hard Bodies which fully touch one another, stick together very strongly. And for explaining how this may be, some have invented hooked Atoms, which is begging the Question' (Newton, *Opticks*, Bk. III, part I).

3 'The only view compatible with this picture [of a brick-like structure for crystals] would be that the single bricks themselves possess these properties [of complete crystals], which does not solve the problem but only pushes it one step farther back' (L. A. Seeber, in Gilbert's *Annal. Phys.* vol. XVI, p. 229).

4 'We may indeed suppose the atom elastic, but this is to endow it with the very property for the explanation of which...the atomic constitution was originally assumed.' It is in questionable scientific taste, after using atoms so freely to get rid of forces acting at sensible distances, to make the whole function of the atoms an action at insensible distances...it merely transfers the difficulty to the primitive atoms' (Clerk Maxwell, 'Atom', in *Scientific Papers*, vol. II, pp. 471, 480).

5 'This attribution of the properties of crystals to the units of which crystals are composed does not solve the problem, but only pushes it one step farther back' (M. von Laue, Introduction, *International Tables for X-ray Crystallography*, vol. I).

6 'Assume matter to be made up of...small constituent parts and... postulate laws for the behaviour of these parts, from which the laws of the matter in bulk could be deduced. This would not complete the explanation, however, since the question of the structure and stability of the constituent parts is left untouched' (Dirac, *Quantum Mechanics*, p. 3). 'It is impossible to explain...qualities of matter except by tracing these back to the behaviour of entities which themselves no longer possess these qualities. If atoms are really to explain the origin of colour and smell of visible material bodies, then they cannot possess properties like colour and smell...atomic theory consistently denies the atom any such perceptible qualities' (Heisenberg, *Die Antike*, vol. VIII). Also cf. 'Should we find that the electron has a complex structure...then such speculation must be pushed one stage farther. ...But even if such a state of affairs were to arise, giving a situation very similar to that which arose when it was discovered the atom had a complex structure, we should merely have pushed the frontier one stage farther back' (G. K. T. Conn, *The Wave Nature of the Electron* (London, 1938), p. 69).

7 Cf. Molière, *Malade Imaginaire* (London, 1732), third ballet scene.

8 I begin here despite Strabo: 'And if one must believe Posidonius, the ancient dogma about atoms originated with Moschus, a Sidonian, born before the Trojan times' (Bk. XVI, II, 24. Loeb Library edition, VII, 271). And in Cudworth: 'That *Ancient Atomick Physiology*... was no Invention of Democritus nor Leucippus, but of much greater Antiquity: not only from that Tradition transmitted by Posidonius the Stoick, but it derived its Original from one Moschus a Phoenician, who lived before the Trojan Wars' (*True Intellectual System*, (London, 1743), the Preface to the Reader).

PAGE 121

1 Boyle put the matter succinctly: 'Matter being in its own nature but one, the diversity we see in bodies must necessarily arise from somewhat else than the matter they consist of' (*Works* (London, 1744), vol. III, p. 15).

2 Democritus, in Diels, *Fragmente der Vorsokratiker* (A 49, A 1, § 45). Cf. also Plato: 'Properties such as hard, warm, and whatever their names may be, are nothing in themselves...' (*Theaetetus*, 156e); also Galileo: 'White or red, bitter or sweet, noisy or silent, fragrant or malodorous, are names for certain effects upon the sense organs' (*Opere*, vol. IV, pp. 333 ff.; cf. also *Two New Sciences*, pp. 40, 48, 112). Cf. also Descartes, *Principia* and *Traité de la Lumière*. Locke, *Enquiry Concerning Human Understanding* (London, 1726), Bk. II, ch. 8, §§ 15–22. Leibniz, *Die Philosophische Schriften*, vol. VII, p. 322; Heisenberg, *Nuclear Physics* (London, 1953), p. 4.

3 Cf. Copernicus: 'Those tiny and indivisible bodies called atoms...are not perceivable by themselves...' (*De Revolutionibus Orbium Cœlestium* (Thoruni, 1873), Bk. I, VI); Lucretius' atoms had no colour; the colour of an aggregate depended on the size and shape of the atoms and their inter-relations (Lucretius, *On the Nature of Things*, trans. Cyril Bailey, Oxford edition, p. 98 (Bk. II, ll. 703 ff.)). The atoms were also without heat, sound, taste or smell (Bk. II, ll. 842 ff., pp. 94 ff.); cf. also Epicurus, *Letter to Herodotus: in Stoic and Epicurean Philosophers* (Modern Library edition, pp. 7–11). Cf. Bacon, 'Bodies entirely even in the particles which affect vision are transparent, bodies simply uneven are white, bodies uneven and in a compound yet regular texture are all colours except black; while bodies uneven and in a compound, irregular, and confused texture are black' (Aphorism xxiii, *Works* (London, 1824), vol. VIII, p. 222.) And Newton: 'The atoms...were themselves, he [Newton] thought, transparent; opacity was caused by "the multitude of reflections caused in their internal parts"' (Birch, *History of the Royal Society*

(London, 1756–7), vol. III, pp. 247 ff.). Stumpf, a good Kantian, found it impossible to imagine the atoms as spatial bodies without colour (cf. *Über den Psychologischen Ursprung der Raumvorstellung* (Leipzig, 1873), p. 22).

PAGE 122

1 Atomists have differed as to which property of the atom, besides indivisibility, was most important. *Position* was paramount for Democritus, *shape* for Epicurus and Lucretius. Newton attended to *motions*, Gassendi remarked the atoms' *combinatory properties*; *irresolvability* attracted Boyle, while for Lavoisier (conservation of mass), Richter (constant proportions) and Dalton, *mass* was the primary consideration; Berzelius (valency) noted their *binding force*, and Prout (again stressing indivisibility) made the hydrogen atoms the ultimate building blocks of matter. Still further properties were stressed by Faraday, Weber, Maxwell, Boltzmann, Clausius, Mayer, Loschmidt and Hittorf. But all atomists used their chosen properties in the same way; to explain the macrophysical properties of bodies by tracing them back to the characteristics of these component particles.

2 Support did not all come from one source. In chemistry, explaining why two elements form a compound without residue only when mixed in discrete mass proportions, gradually becomes 'atomic'. (If elements consist of particles of definite mass, and if forming a compound involves grouping these, then the laws of definite and multiple proportions follow as a matter of course from the theory of the atomic structure of matter.)

Faraday's approach was from another direction. He writes: 'Equivalent weights of bodies, are simply those quantities of them which contain equal quantities of electricity....Or, if we adopt the atomic theory...then the atoms of bodies...have equal quantities of electricity naturally associated with them.' For Helmholtz too (1881) Faraday's laws of electrolysis suggested the existence of atoms of electricity.

3 Ostwald, Mach and Pearson were strict. Ostwald banished the concept *atom* from physical science. Mach wrote that 'The atomic theory is [merely] a *mathematical model* used for the representation of facts'. Also, 'Atoms, electrons, and quanta are only links (auxiliary concepts) to represent a connected system of science' (*Mechanics*, pp. 590 ff.). Vaihinger argues: 'Simple atoms [entities without extension]...cannot be actual things' (*Die Philosophie der 'Als Ob'* (Berlin, 1932), p. 219). Pearson puts it that 'Atom and molecule are *intellectual conceptions* by aid of which physicists classify phenomena' (*Grammar of Science* (London, 1911), p. 85). Cassirer: 'A concept like that of material point...can never be understood as the copy of a physical

object' (*Determinismus und Indeterminismus in der Modernen Physik*, pp. 164ff.).

4 H. Nagaoka, *Nature*, LXIX, (1904), 392. Rutherford, 'The Scattering of α and β particles by Matter', *London, Edinburgh and Dublin Philosophical Magazine* (1911), p. 688. Bohr's theory of the H-atom was a space-time atomic model involving a mechanical orbit for the particle.

5 'I...suspect that [the phenomena of Nature] may all depend upon certain forces by which the particles of bodies...are either mutually impelled towards one another and cohere in regular figures, or are repelled and recede from one another' (Newton, *Principia*, p. xviii).

6 Cf. '[The Democritean atoms are] *by hypothesis* the result of division to the last possible stage...hence they cannot in themselves undergo any of the changes experienced by sense-perception' (K. Freeman, *Ancilla to the Pre-Socratic Philosophers*, Oxford, Blackwell, 1948).

PAGE 123

1 Cf. De Broglie, *The Revolution in Physics* (London, 1954), p. 33. Dirac, *Quantum Mechanics*, pp. 1, 2, 35. W. Heitler, *Elementary Wave Mechanics* (Oxford, 1945), p. 1, ll. 1–7.

2 'The true atoms (in the sense of the ancients) are the elementary corpuscles, for example electrons, which today are considered (perhaps tentatively) as the ultimate constituents of atoms and hence of matter' (De Broglie, *The Revolution in Physics*, p. 60).

3 Cf. Dirac, *Quantum Mechanics*, pp. 9, 14.

4 Other methods are the simple ionization chamber, the point counter (Geiger), the tracer (for neutrons), and, more recently, the bubble chamber. 'The existence and properties of the ultimate elements are only to be inferred indirectly from observations of gross matter, e.g....as in Millikan's experiments' (G. Temple, *The General Principles of Quantum Theory* (London, 1934), p. 23). Cf. C. T. R. Wilson's important paper, *Proc. R. Soc.* A, LXXXV (1911), 285.

5 'After a study of these [α-particle tracks] one can no longer have the slightest doubt that very small particles have actually flown through space' (Heisenberg, *Nuclear Physics*, p. 33). Cf. R. A. Millikan, *Electrons (+ and −), Protons, Photons, Neutrons, and Cosmic Rays* (Cambridge, 1935) an *opus classicus* of the particulate nature of the electron.

The wave nature of the electron was established by C. Davisson and L. H. Germer, *Phys. Rev.* XXX (1927), 707; G. P. Thomson, *Proc. R. Soc.* A, CXVII (1928), 600; Rupp, *Ann. Phys., Lpz.*, LXXXV (1928), 901; Kikuchi, *Jap. J. Phys.* V (1928), 83; Ponte, *Comptes*

Rendus, CLXXXVIII (1929). The wave nature of particles of atomic mass was established by F. Knauer and O. Stern, *Z. Phys.* LIII (1929), 786; I. Estermann and Stern, *ibid.* LXI (1930), 115. Cf. also T. H. Johnson, *J. Franklin Inst.* CCVI (1928), 301; Ellett, Olson and Zahl, *Phys. Rev.* XXXIV (1929), 493.

6 *Philosophical* uneasiness, about, e.g. the uncertainty relations, is no good reason to doubt that electrons do have properties which entail these relations. Einstein, Bohm, De Broglie are just wrong; there *are* shortcomings in quantum concepts, but these concern inconsistencies, not hunches that God is not a dice-player.

7 '[Quantum mechanics] requires the states of a dynamical system and the dynamical variables to be interconnected in quite strange ways that are unintelligible from the classical standpoint' (Dirac, *Quantum Mechanics*, p. 15).

PAGE 124

1 This is the retroduction discussed in ch. IV. Examples follow: 'Similar effects [to what is observed with water waves] take place whenever two portions of light are thus mixed; and this I explain by the general law of the interference of light' (Young, *Works* (London, 1855), vol. I, p. 202); and Rutherford: 'Considering the evidence as a whole, it seems simplest to suppose that the atom contains a central charge distributed through a very small volume' ('The Scattering of α and β Particles by Matter', *London, Edinburgh and Dublin Philosophical Magazine* (1911), p. 687); and again in 1919: 'It is difficult to avoid the conclusion that the long-range atoms arising from collision of α particles with nitrogen, are not nitrogen atoms, but probably atoms of hydrogen, or atoms of mass 2' ('Collision of α Particles with Light Atoms', *ibid.* (1919), p. 581); 'It was known that many spectral lines consisted of two very close lying components....In 1925 this doubling was explained [by Goudsmit and Uhlenbeck] in terms of a new property of the electron, *its spin*' (Humphreys and Beringer, *Atomic Physics*, p. 279); and Dirac: 'We are led to infer that the negative-energy solutions of [equation 56, *Quantum Mechanics*, p. 272] refer to the motion of a new kind of particle having the mass of an electron and the opposite charge [the positron]' (*ibid.* (1930), p. 273); and Anderson: 'The tracks shown in Fig. 1 were obtained, which seemed to be interpretable only on the basis of the existence in this case of a particle carrying a positive charge, but having the mass of...a free negative electron' ('The Positive Electron', *Phys. Rev.* (1933), p. 491); and Chadwick: 'The experimental results were very difficult to explain on the hypothesis that the beryllium radiation was a quantum radiation, but followed immediately if it were supposed that the radiation consisted of particles of mass nearly equal to that of a proton and with no net charge, or neutrons.' 'The simplest hypothesis one can make about the nature of the particle is to suppose that it consists of a proton and an electron in close combination, giving a net charge 0 and a mass which should be slightly less than the mass of the hydrogen atom' ('Existence of a Neutron', *Proc. R. Soc.* (1932), pp. 694, 700). [Carrying no charge, the neutron leaves no track in a cloud chamber. This made its detection (by means of a boron counter) difficult, and insured that evidence of it would be circumstantial, i.e. retroductive]; Yukawa: 'The interactions of elementary particles are described by considering a hypothetical quantum which has the elementary charge and the proper mass and which obeys Bose's statistics. The interaction of such a quantum with the heavy particle should be far greater than that with the light particle in order to account for the large interaction of the neutron and the proton as well as the small probability ot β disintegration. Such quanta [later called "mesons"], if they ever exist and approach the matter close enough to be absorbed, will deliver their charge and energy to the latter.... The massive quanta may also have some bearing on the shower produced by cosmic rays...' ('On the Interaction of Elementary Particles', *Proc. Physico-Mathematical Society of Japan* (1935)). [It was Anderson again who experimentally detected the meson (the heavy, or *tau* meson) (*Phys. Rev.* LI (1937), 884)]; and Fermi: 'The conservation of momentum in the case of the electromagnetic interaction is a necessary condition in order to have a non-vanishing matrix element' (*Elementary Particles*, p. 20); and Frisch: 'For neither of those assumptions [of the independent-particle model of the atomic nucleus] a good argument could (or can) be made. But once you accept them, a number of dimly noticed regularities fall into place; others were foretold and duly verified' (*Cambridge Rev.* (1955)).

Cf. 'The wave field of an electron...is a probability [amplitude]... without the ψ function no laws of nature could be formulated in the domain of atomic physics' (Heitler, *Elementary Wave Mechanics*, p. 14). 'Its [ψ] role is mathematical, but for that reason none the less vitally important for the description of nature' (*ibid.* p. 42).

'ψ...is linked up with physically observable quantities through a rather long chain of abstractions' (O. Halpern and H. Thirring, *The New Quantum Mechanics* (London, 1932), p. 43).

Retroductions do not always lead to syntheses like those of Newton, Clerk Maxwell, Einstein and Dirac. They sometimes show the first chink in the old armour, as when the good Newtonian Leverrier 'explained' Mercury's precessions by the non-existent planet Vulcan, a move which had proved monumentally successful in the case of Uranus and Neptune.

2 E.g. to explain surprising observations in series spectra and the anomalous Zeeman effect, Uhlenbeck and Goudsmit (*Naturwissenschaften*, XIII (1925), 953), suggested that the electron was a spinning particle. This proposal was built by Dirac (*Proc. Lond. Phys. Soc.* CXVII, 610; CXVIII, 351 (1928)) into a relativistically invariant theory, yielding spin properties without additional assumptions, and explaining Sommerfeld's fine-structure formula (series spectra) and the separations and intensities in the Zeeman effect (doublet atoms). Spin he now regarded as a fundamental property, a premise of the entire theory, not an accidental adjunct.

3 Fermi, *Elementary Particles* (New Haven, 1951), p. 2.

PAGE 125

1 Cf. the fundamental paper of Fermi, 'Versuch Einer Theorie der β-Strahlen', *Z. Phys.* LXXXVIII (1934), 161. Cf. Heisenberg, *Nuclear Physics*, pp. 122–4.

The actual experimental discovery of the neutrino is reported by Dr Reines and Dr Cowan, both of Los Alamos ('The Neutrino', *Nature* (1956)). Their work is impressive, even if certain eminent experimentalists still retain doubts about whether the particle has in fact been isolated. Cf. also H. R. Crane, *Rev. Mod. Phys.* XX (1948), 278, which summarizes all neutrino detection attempts up to 1948.

PAGE 126

1 Cf. Heisenberg, *Philosophical Problems of Nuclear Science* (London, 1952), pp. 101, 105. Cf. Maxwell writing in 1855: 'The student must make himself familiar with a considerable body of most intricate mathematics.... The first process therefore... must be one of simplification and reduction of the results of previous investigation to a form in which the mind can grasp them. The results of this simplification may take the form of a purely mathematical formula' ('On Faraday's Lines of Force', *Trans. Camb. Phil. Soc.* X, no. 1).

2 It is like according to the limit of a series, properties of the series' members: a notoriously unsound procedure.

Cf. 'There is an entirely new idea involved, to which one must get accustomed, and in terms of which one must proceed to build up an exact mathematical theory, without having any detailed classical picture' (Dirac, *Quantum Mechanics*, p. 12). 'The main object of physical science is not the provision of pictures, but is the formulation of laws governing phenomena and the application of these laws to the discovery of new phenomena... whether a picture exists or not is a matter of only secondary importance. In the case of atomic phenomena no picture can be expected to exist in the usual sense of the word "picture", by which is meant a model functioning essentially on classical lines' (*ibid.* p. 10). 'To interpret nature on engineering lines proved equally inadequate...to interpret nature in terms of the concepts of pure mathematics [is]...brilliantly successful' (Jeans, *The Mysterious Universe* (Cambridge, 1930), p. 143). '*All* the pictures which science now draws of nature, and which alone seem capable of according with observational fact, are *mathematical* pictures'....'The universe begins to look more like a great thought than like a great machine' (*ibid.* pp. 135, 143). 'The truly creative principle (for physics) resides in mathematics' (Einstein, *Herbert Spencer Lecture* (1933)). 'We have reached the limits of visualization...the concept of electrons circling a nucleus cannot be taken literally' (Heisenberg, *Nuclear Physics*, p. 30).

'This picture of the spinning electron as a rotating ball must not be taken literally. No physical reality whatsoever can be attached to the "structure of the electron"...questions of what the "radius" of such a ball would be, etc., are void of any physical meaning' (Heitler, *Elementary Wave Mechanics*, p. 70). 'If we persist in describing phenomena according to the methods of classical physics by means of space and time, then we must give up our ideas of continuity....If we wish to retain...continuity...we must give up space-time description....We must not expect to be able easily to picture by means of models the fundamental things of nature' (Flint, *Wave Mechanics* (London, 1951), p. 110). 'Schematic idealizations [pictures, classical models]...are capable of representing certain aspects of things, but they have their limits and cannot incorporate into their rigid forms all the richness of reality' (De Broglie, *The Revolution in Physics*, p. 19). 'It is very difficult to modify our language so that it will be able to describe these atomic processes, for words can only describe things of which we can form mental pictures, and this ability, too, is a result of daily experience. Fortunately, mathematics is not subject to this limitation, and it has been possible to invent a mathematical scheme —the quantum theory—which seems entirely adequate for the treatment of atomic processes' (Heisenberg, *The Physical Principles of the Quantum Theory* (Chicago, 1930), p. 11).

PAGE 127

1 This is what Einstein did with light (*Ann. Phys.* IV (1905), 17). Because it throws sharp shadows and does not bend around corners, Newton likened light to a stream of particles. Huygens, Young, Fresnel and Foucault discredited this, shaping our classical ideas of light along undulatory lines. Einstein demurred, explaining the photoelectric effect by quanta of radiation modelled on Planck's researches. Although J. J. Thomson's experiments had induced Lorentz to bring discontinuity into Maxwell's theory, the contradictoriness of this did

not stand out until Einstein explicitly joined particulate and undulatory properties in his 1905 paper.

Physicists of the time thought the 'dual nature' of light a paradox; against a nineteenth-century pattern of concepts it was. It was 'the surprising phenomenon' whose explanation was nothing less than the whole quantum theory. See, e.g. Sommerfeld, *Atombau und Spektrallinien* (London, 1923); Reiche, *The Quantum Theory* (London, 1922); Landé, *Fortschritte der Quantentheorie* (Dresden, 1922). Einstein prepared the way for Compton's theory of matter-scattered photons (*Phys. Rev.* XXI (1923), 483; and (with Simon), *ibid.* XXVI (1925). Cf. Bethe, 'Quanta', *Handbuch der Physik*, XXIII (1926), iii, § 73). Without this lead the De Broglie and Schrödinger equations could hardly have been constructed. Cf. Schrödinger, *Wave Mechanics* (London, 1928), p. 5; von Neumann, *Mathematical Foundations of Quantum Mechanics* (Princeton, 1955), pp. 3, 212; Halpern and Thirring, *The New Quantum Mechanics*, pp. 137ff.; and Heitler, *Elementary Wave Mechanics*, 'The velocity v [in De Broglie's $\lambda = h/mv$] is a concept relevant to the electron pictured as a particle, whilst λ is a concept relevant to a wave' (p. 4).

For a detailed discussion of the photo-electric effect, see Mott and Sneddon, *Wave Mechanics and its Applications* (Oxford, 1948), p. 139; Heitler, *Quantum Theory of Radiation* (Oxford, 1936), p. 122; Bethe, *Handbuch der Physik*, XXIV (1933), 475; Stobbe, *Ann. Phys.* VII (1930), 661.

PAGE 131

1 Notice how this compares with Bohr's quantum conditions, $I_k = n_k h$, *Philosophical Magazine*, XXVI (1913); *Z. Phys.* VI (1920). Cf. also Hertz, *Verh. dtsch. phys. Ges.* XV (1913). Cf. von Neumann, *op. cit.* p. 287 and n. 150.

PAGE 132

1 This comes from Sommerfeld, *Atombau und Spektrallinien.*

PAGE 133

1 'The exchange charge and the exchange energy is...due to the fact that two electrons are indistinguishable from each other' (Heitler, *Elementary Wave Mechanics*, p. 80). 'The light quanta have now the fundamental property of being exactly identical, i.e. there is no conceivable way of distinguishing between two light quanta with the same co-ordinate' (von Neumann, *Mathematical Foundations*, p. 275). 'One cannot speak of *the* electron, but merely of *an* electron' (Conn, *The Wave Nature of the Electron*, p. vi).

The Pauli-Fermi argument for the existence of a neutrino appeals to the Identity Principle. The α-particles produced by a homogeneous radioactive substance are required to be identical; they have exactly the same range and hence the same energy. [The only theory which gives us any adequate concept of this particle describes each α-ray with the same parameters in the same notation. How *could* they differ? If you persist, saying they could differ in some undetected way, what are you advancing for consideration? If there were a difference between α-particles with respect to a property dealt with in Fermi's theory, then that would prove that the particles originated from a non-homogeneous substance, i.e. from different kinds of nuclei. But if the 'difference' suggested has to do with some property not dealt with in Fermi's theory, then either you must have a better alternative theory, or you literally have no idea of what you are talking about (cf. ch. II).]

β-particles, however, are emitted with all possible velocities. This, if unexplained, would appear to wreck the energy principle, the identity principle, and elementary particle physics; but the continuous β-ray spectrum is taken rather to prove the existence of another particle, the neutrino—whose energy is always just what is required to bring the total energy of the β-particle-neutrino 'pair' up to the maximum value for any given homogeneous substance. This removes the energy principle (and hence all physics) from danger.

Note that the identity principle, which is on such intimate terms here with the energy principle, was also central to the example which concluded ch. IV. There an event which must be completely 'uncaused' (in the classical sense) was remarked. For every C_{14} nucleus must be identical with every other one right up until the time when some of them decay radioactively. However, it is impossible in principle to predict which ones will decay. Cf. von Neumann, *Mathematical Foundations*, pp. 206, 207.

PAGE 136

1 Heisenberg, 'Über den anschaulichen Inhalt der quantentheoretischen Kinematik und Mechanik', in *Z. Phys.* XLIII (1927), 172. Heisenberg's reflexions were extended by Bohr (*Naturwissenschaften*, XVI (1928)).

2 Heisenberg, *op. cit.* Bohr, *Nature*, CXXI (1928), 580. *Naturwissenschaften*, XVI (1928), 245; but esp. *Atomtheorie und Naturbeschreibung* (Berlin, 1931). This example is discussed in every standard work on quantum theory and wave mechanics. Cf. particularly Heitler, *The Quantum Theory of Radiation*, pp. 56ff. Cf. also N. R. Hanson, 'Uncertainty', *Phil. Rev.* (1954).

PAGE 137

1 'It is impossible, with light of wave-length λ, to picture sharply objects which are smaller than λ, or even to reduce the scattering to such an

extent that one can speak of a (distorted) image' (von Neumann, *Mathematical Foundations*, p. 240).

Cf. also Conn, *The Wave Nature of the Electron*, p. 62, and Flint, *op. cit.* p. 82. For the theory of the microscope see *Handbuch der Physik* (Berlin 1927), vol. 18, ch. 2 G. And cf. the interesting remark of Heisenberg (*The Physical Principles of the Quantum Theory*, p. 34) on the feasibility of using red light.

2 A normal lens fails with γ-rays; its molecules would be perturbed or shattered by the radiation. One needs a different kind of focusing device. Cf. L. Marton, 'A New Electron Microscope', *Phys. Rev.* LVIII (1940), p. 57; also Bachman and Ramo, 'Electrostatic Electron Microscopy', *J. Appl. Phys.* XIV, 8–18, 69–72, 155–60.

3 'Each conceivable measuring apparatus, as a consequence of the imperfections of human means of observation...can furnish this value only with a certain (never vanishing) margin of error' (von Neumann, *Mathematical Foundations*, p. 221).

PAGE 138

1 See the classical paper: Friedrich, Knipping and von Laue, 'Interference Phenomena with Röntgen Rays' (K. Bayer, *Akad. München* (Berlin, 1912), pp. 303–22).

2 'This impossibility is not due to any shortcoming (still remediable) of that postulated ideal microscope, assumed to be as perfect as natural laws would allow it to be, but rather a consequence of those very laws' (Heisenberg, *Nuclear Physics*, p. 29). 'From the simplest laws of optics, together with the empirically established law $\lambda = h/p$, it can be readily shown that $\Delta x \Delta p_x \geqq h$' (Heisenberg, *The Physical Principles of the Quantum Theory*, p. 14). 'The well known laws of wave optics, electrodynamics and elementary atomic processes, place very great difficulties in the way of accurate measurement precisely where this is required [to illustrate the physical significance of] the uncertainty relations' (von Neumann, *Mathematical Foundations*, p. 238).

PAGE 140

1 Von Neumann remarks the basic difficulty even in defining a (classical) electric field, namely, that the electrical test charge to be used cannot be smaller than an electron (*op. cit.* p. 300, note 159). Temple (*The Principles of Quantum Theory*, p. 23) puts the conceptual kernel of this into a nutshell: 'The system measured must be at the same time an isolated whole and a part interacting with other parts. Measurement is impossible unless the system acts upon the apparatus of observation, and the measurements are meaningless unless the system retains its identity and characteristics.'

An excellent discussion of the concept of the *Probekörper* is to be found on pp. 192–3 of Halpern and Thirring, *The New Quantum Mechanics*.

PAGE 141

1 D. Bohm (*Phys. Rev.* LXXXV (1952)), whose ideas are a serious challenge to this statement, will be dealt with later in an Appendix.

2 The uncertainty principle is not a theoretical summary of data (e.g. the photo-electric and Compton effects, cathode- and β-ray diffraction, etc.). Rather, it is in order to explain these data that the concepts which entail the uncertainty relations must be retroductively inferred. Von Neumann (*Mathematical Foundations*, p. 295) generates all of quantum theory from a formula which contains the uncertainty relations in a very obvious way. Bohm writes: 'The uncertainty principle is...a necessary consequence of the assumption that the wave function and its probability interpretation provide the most complete possible specification of the state of an individual system' (*Phys. Rev.* LXXXV (1952), 167). And Heitler puts it that 'The quantum properties of a particle are contained in the uncertainty relations for the position and the momentum' (*The Quantum Theory of Radiation*, p. 57). In other words, quantum theory can as easily be generated from the uncertainty relations, as *vice versa*. This is rarely appreciated by the theory's critics. The uncertainty relations are more than just an unavoidable consequence of the theory. In a sense they *are* the theory. Thus Heisenberg (*Nuclear Physics*, p. 93) infers from the diameter of the deuteron, *in accordance with the uncertainty principle*, to the average magnitude of the kinetic energy of a proton-neutron system.

3 'By a particle surely we understand a small body of definite size which at any definite instant of time has some definite position, that is, occupies some specific portion of space' (Conn, *The Wave Nature of the Electron*, p. 61).

4 'The characteristic properties of a wave propagation are that the distribution of energy in space should, at the detector, display a certain periodic pattern, which is not only *not* characteristic of a particle propagation, but even inconsistent with it' (*ibid.* p. 19).

PAGE 142

1 Thus von Neumann speaks of 'The self-contradictory dual nature of light' (*Mathematical Foundations*, p. 4). This consequence of De Broglie's matter-wave hypothesis was regarded by many physicists thirty years ago as proof that the allocation of a vibration event to a material particle was impossible and senseless. Cf. Halpern and Thirring, *op. cit.* p. 36, and Conn, *op. cit.* p. 46.

2 After all, Mach notes how casual we are with 'bevel-edged cubes' and 'perforated cylinders' (*Mechanics*).

PAGE 143

1 'According to Bohr, this restriction may be deduced from the principle that the processes of atomic physics can be visualized equally well in terms of waves or particles' (Heisenberg, *The Physical Principles of the Quantum Theory*, p. 13).

PAGE 144

1 'The two concepts [v and λ] are connected by Planck's constant h' (Heitler, *Elementary Wave Mechanics*, p. 4).

2 De Broglie, *Thèses* (Paris, 1924); *Phil. Mag.* XCVII (1924); *Annales de Physique*, X (1925), 22.

It should be noted that Einstein (*Berlin Ber.* (1924), 261; III (1925), 18) had an independently formed conception of material waves, of which he made use in these papers dealing with the statistics of an ideal gas. Schrödinger's inspiration, however, came from De Broglie. Note, however, that for De Broglie material particles are *associated* with waves; for Schrödinger particles *are* waves.

3 The idea of co-ordinating optical concepts with purely mechanical ones by localizing the former is very ancient. Qualitative suggestions are also to be found in Newton's *Opticks* (Bk. III). The first rigorous treatment, of course, is that of Fermat. Hamilton stressed analogies between these two branches of physics, but until De Broglie and Schrödinger little had been done with them, except perhaps by Fourier.

4 These are De Broglie's *ondes de phase* (*Thèses*).

PAGE 145

1 These figures are set out in Flamm, 'Die neue Mechanik', *Naturwissenschaften*, XV (1927), 569. Cf. Halpern and Thirring, *The New Quantum Mechanics*, § 67.

One of our main conclusions is already built into the diagram. A wave-pulse consisting of several monochromatic waves cannot, *logically cannot*, have a sharp momentum, since if a wave-function ψ consists of several parts $\Sigma_k \psi_k$ such that for each part Q has a sharp value (but the qk's are different) then Q cannot be sharp for a state described by the wave-function.

2 E. Schrödinger, *Ann. Phys.* LXXIX (1926), 361 and 489; cf. also *Four Lectures on Wave Mechanics* (London, 1928); cf. Conn, *The Wave Nature of the Electron*; Heitler, *Elementary Wave Mechanics*, and also his *Quantum Theory of Radiation*; Flint, *Wave Mechanics*; and Mott, *Elementary Wave Mechanics*.

PAGE 146

1 'Classical' because the Schrödinger formulation is our starting-point. Schrödinger's is a classical theory, based completely on ideas of continuity. He aspires to link wave mechanics with classical field theory (cf. *Four Lectures on Wave Mechanics*, pp. 5, 6; and Halpern and Thirring, *op. cit.* p. 183). Had we begun with the experimentally more serviceable views of Bohr, Dirac, Jordan, Born and Heisenberg —all of whom based their theories on a non-classical concept of discontinuity—this exposition would not have proceeded as it has. More will be said in Appendix II about this fundamental philosophical difference between wave mechanics and quantum mechanics. We may anticipate this with Boltzmann's observation: 'The question is whether the pure differential equations or atomism will one day turn out the more complete descriptions of phenomena' (*Vorlesungen über Gastheorie* (Leipzig, 1896–8), vol. I, p. 6). De Broglie, Schrödinger, Flint, Conn, etc., view wave mechanics as resting on the pure differential equations. Bohr, Heisenberg, Jordan, Born, Dirac—and now almost all other physicists—view quantum mechanics as resting on 'atomism'. Yet wave mechanics and quantum mechanics are operationally equivalent, 'at least in a mathematical sense' (von Neumann, *Mathematical Foundations*, p. 5). This is an instructive analogue of the situation wherein Kepler and Tycho are watching the dawn. The conceptual difference between these two observationally equivalent physical languages is enormous, just as with our earlier examples.

PAGE 148

1 This situation is set out in detail by Halpern and Thirring, *op. cit.* § 70, where the interacting parameters are as they had been originally for Heisenberg (1927), the particle's energy and its time-behaviour. Bohr (1928) found the same relations for the specific pair: position and momentum. Von Neumann (*op. cit.*) generalizes this for any pair of non-commuting operators (pp. 229, 234, 251: 'E, F are simultaneously decidable if and only if the corresponding quantities E, F are simultaneously measurable...i.e. if [operators] E and F commute'). This will be developed in the next section.

2 Wave mechanics and matrix mechanics were shown to be experimentally equivalent by Schrödinger, *Annal. Phys.* LXXIX (1926), 734 and Eckart, *Phys. Rev.* XXVIII (1926), 711. Dirac (*Proc. R. Soc.* CIX (1925); CXIII (1926)) and Jordan (*Z. Phys.* XL (1926)) welded the two notations into the powerful formalism so strikingly set out in Dirac's *Quantum Mechanics* (1930). 'The uncertainty relations can also be deduced without explicit use of the wave picture, for they are readily obtained from the mathematical scheme of quantum theory' (Heisenberg, *The*

Physical Principles of the Quantum Theory, p. 15). Cf. Kennard, *Z. Phys.* XLIV (1927), p. 326.

3 'One obtains the limitations of the concept of a particle by considering the concept of a wave...one may derive the limitations of the concept of a wave by comparison with the concept of a particle' (Heisenberg, *op. cit.* p. 11).

PAGE 149

1 'An introduction of hidden parameters is certainly not possible without a basic change in the present theory' (von Neumann, *op. cit.* p. 210).

2 'Quantum mechanics has, in its present form, several serious lacunae, and it may even be that it is false' (*ibid.* p. 327). Most physicists realize that there is a lot of work to be done in elementary particle theory as it stands (cf. Hanson, 'On Elementary Particle Theory', in *Philosophy of Science* (1956): 'Sur la théorie des particules élémentaires'; *Scientia* (1956)); but it can be said emphatically that there is nothing in this to herald a new 'deterministic' theory, despite De Broglie's enthusiasm (cf. *The Revolution in Physics*, p. 302).

3 Weyl, *Philosophy of Mathematics and Natural Science*, pp. 185–6.

4 Theoretical works (of which Weyl's *Gruppentheorie und Quantenmechanik* (Leipzig, 1928) is one), appeal to the correspondence principle in at least two ways, both legitimate and clear. (1) In the formative days, namely, 1913–27, quantum physicists often used classical mechanics as a criterion for the correctness of their calculations, and as a storehouse of suggestions about research and development within the theory. Bohr continually appealed to the Principle for these purposes (cf. Conn, *The Wave Nature of the Electron*, p. 37). One of his associates worked out a complete quantum theoretical account of the Stark effect on the basis of this principle. [Kramers, *Dissertation* (*Kgl. Danske Vidensk. Selsk. Skrifter*, *Naturvidensk.* Afd. 8, Raekke), iii, III (1919), 287. Cf. also Kramers and Heisenberg, *Z. Phys.* XXXI (1925) where the correspondence principle is used for dispersion problems.] Schrödinger and Dirac have often looked to classical physics for new ideas (cf. Dirac, *Quantum Mechanics*, pp. 84, 85). (2) It is a standard in all science that whatever the other merits of a new theory, unless it explains everything that could be explained by the theory it purports to replace, it is a non-starter (cf. Mott, *Elements of Wave Mechanics*, p. 26 (i)). Many other ideas have succeeded largely because they comply with this conceptual condition (cf. Fermi, *Elementary Particles*, pp. 104–5, and Halpern and Thirring, *The New Quantum Mechanics*, p. 201). Weyl's remark, however,

besides confounding these legitimate appeals, stirs in some pseudo-philosophical propaganda about the comparative merits of the 'old' and the 'new' physics.

Cf. Temple: 'The [correspondence principle] is that the characteristics of microphysical systems are expressed by variables of a type similar to those which describe macrophysical systems, i.e. by a set of positional co-ordinates, together with their associated momenta... this assumption is very simple and plausible, but it definitely transcends our empirical knowledge, and is independent of the general laws of microphysical measurement' (*The General Principles of Quantum Theory*, p. 44). Cf. Heitler, *The Quantum Theory of Radiation*: 'For the problem of emission and absorption of light, the quantum theory gives results which correspond in every detail to those of the classical theory, in the sense of Bohr's correspondence principle' (p. 110).

PAGE 150

1 Nor in Heisenberg's (matrices), Bohr's (q-numbers), Born's (statistical), Wiener's, Schrödinger's, Jordan's or von Neumann's (all operator calculi). Cf. 'We know nothing of the possibility of commutating the co-ordinate matrix with the momentum matrix...the following relation must hold for them: $pq-qp=(h/2\pi i)\,1$' (Halpern and Thirring, *op. cit.* p. 29). 'Only those quantities can be sharply defined simultaneously that are commutative (in the sense of the calculus of operators)...the mathematical criterion that a function should contain the one quantity of conjugate to a quantity p is that it should be non-commutative, that is $(fq)\psi \neq (qf)\psi$...' (*ibid.* p. 197). Cf. also § 29.

See especially the important section (II, 10) in von Neumann, where the logic of commutative operators is discussed: '$PQ-QP$ need not have sense everywhere' (p. 234). 'If p, q are two canonically conjugate quantities and a system is in a state in which the value of p can be given with the accuracy ϵ...then q can be known with no greater accuracy than $\eta=h/2\pi:\epsilon$...' (p. 238) [this follows 'mathematically' (p. 233) from the specification of P and Q as 'non-commutative operators']. And see n. 164, where Heisenberg's characterization of the electron in terms of two non-commuting operators is shown to be such that the possibility of the particle being a punctiform mass is either meaningless in Heisenberg's notation, or it constitutes a recommendation that the entire notation be rejected. Bold physicists have made this recommendation, but never in a persuasive way; this applies even to Bohm. See further: 'The characteristic condition for the simultaneous measurability of an arbitrary (finite) number of quantities...is the commutativity of their operators' (von Neumann, *Mathematical Foundations*, p. 229).

'These quantum-mechanical operators M_z, M_y, M_x *cannot be commuted with one another*....Hence there is no meaning in saying that we can...simultaneously make measurements of M_z and M_y... we cannot regard it as a paradox that the maximum value of M_z^2 does not coincide with M^2 but rather as a beautiful and striking confirmation of the line of reasoning on which Heisenberg's Uncertainty Relation is founded' (Halpern and Thirring, *op. cit.* p. 205). 'If ξ and η are two observables such that their simultaneous eigenstates form a complete set, then ξ and η commute....When the two observables commute, the observations are to be considered as non-interfering or compatible' (Dirac, *Quantum Mechanics*, pp. 49, 52); cf. also § 25. 'One of the dominant features of this scheme is that observables, and dynamical variables in general, appear in it as quantities which do not obey the commutative law of multiplication' (p. 84). And, finally: 'Heisenberg's Principle...shows clearly the limitations in the possibility of simultaneously assigning numerical values, for any particular state, to two non-commuting observables...and provides a plain illustration of how observations in quantum mechanics may be incompatible' (p. 98). Cf. also 'Commutation and uncertainty relations of the field strengths' (§ 8 of Heitler, *The Quantum Theory of Radiation*).

PAGE 151

1 More strongly, not one of the standard notations for elementary particle physics are such that they can be made to express

$$d/dt(mv_x) = X = -(\partial V/\partial x),$$
$$d/dt(mv_y) = Y = -(\partial V/\partial y),$$
$$d/dt(mv_z) = Z = -(\partial V/\partial z),$$

[where force components X, Y and Z have a potential V].

These fundamental equations of classical mechanics are *not even formulable* in the language-games of elementary particle physics; the symbols do not fit together in this way (cf. following note, and recall ch. II). Is it suggested that they might one day be so fitted together, that this limitation is only temporary? Just what substitute concept is one being invited to entertain? There is no idea of an electron (other than that of quantum physics) which cannot be demolished by one of the simple observations that the wave-particle *explicans* was developed to deal with. So what alternative *is* there to the uncertainty relations which this *explicans* entails? Absolutely none. See the excellent discussion on p. 43 of Heitler's *Elementary Wave Mechanics*.

2 Another philosophical aspect of Weyl's remark concerns whether it is really true that the relation of quantum mechanics to classical dynamics is similar to the relation of the latter to relativity physics.

There are reasons for denying this; the two situations are logically quite distinct. But the discussion cannot be pursued here. Von Neumann remarks the matter (*op. cit.* pp. 325, 326). Cf. also Heisenberg, *The Physical Principles of the Quantum Theory*, p. 2.

PAGE 152

1 'p and q are two incompatible variables...homogeneity with respect to one variable, say q, implies an infinite uncertainty in any incompatible variable p' (Temple, *The General Principles of Quantum Theory*, p. 43).

2 'Classical mechanics must therefore be a limiting case of quantum mechanics' (Dirac, *Quantum Mechanics*, p. 84).

3 Quantum theory is a language, with its own formation rules and transformation rules. Why not say that a cluster of symbols which breaks its rules, as does, e.g. '$d^2s/dt^2 = \ddot{v}$', expresses no intelligible assertion in the language? Thus: 'All the well-known but not understood "rules" come out one after the other as the result of...absolutely cogent analysis...once the hypothesis about $\psi\bar{\psi}$ has been made, no accessory hypothesis is needed or is possible' (Schrödinger, *Wave Mechanics*, p. 20). And cf. the important passage in Heisenberg: 'Any use of the words "position" and "velocity" with an accuracy exceeding that given by equation [$\Delta p . \Delta v \simeq h/m$] is just as meaningless as the use of words whose sense is not defined' (*Principles of the Quantum Theory*, p. 15).

4 'Up to [1925]...quantum theory...was a conglomeration of essentially different...and partially contradictory fragments, [e.g.] the correspondence principle, belonging half to classical mechanics and electrodynamics [and]...the self-contradictory dual nature of light' (von Neumann, *op. cit.* p. 4). Not every trace of these origins was obliterated by the Dirac-von Neumann synthesis, as Weyl's remark reveals.

5 Professor Sir Harold Jeffreys assures me that J. E. Moyal has surmounted this limitation in the statistical formulation (cf. *Proc. Camb. Phil. Soc.* XLV (1948), 99–124).

Sir Harold also writes: 'The whole of classical mechanics depends on the existence of such [simultaneous probability] distributions. If we knew only the position of a body at an instant, and nothing at all about its momentum, we could predict nothing at all about its position at any other instant. Quantum mechanics, if it is to be comprehensive, must be in a position to derive the classical equations of motion as approximations valid for systems containing many atoms; and, however this is done, some variables corresponding to the co-ordinates and momenta must persist. To deny that they can have a simultaneous probability distribution is to say that quantum mechanics can never explain why classical mechanics gives the right answers for the motion of the planets' (*Scientific Inference*, Cambridge, 2nd ed. 1957, p. 218).

PAGE 153

1 From a private communication by Sir Harold Jeffreys. It appears here with his permission. Von Neumann considers this possibility: 'The most obvious first step [after encountering the limitations in the theory] would be to assume that this is an incompleteness...that there must exist a more general formula embracing this as a special case' (*Mathematical Foundations*, p. 211).

2 Von Neumann continues: 'Such a generalization...is not possible... in addition to the formal reasons (intrinsic in the structure of the mathematical tools of the theory) weighty physical grounds also suggest this type of limitation' (*ibid. loc. cit.*). 'Everything which can be said about the state of a system must be derived from its wave function' (p. 196). 'A discontinuous operator can never be made continuous by extension' (p. 149). Cf. Heitler, *Elementary Wave Mechanics*, p. 10.

PAGE 154

1 Cf. again Heitler, *ibid.* p. 18, ll. 7–16.

2 '$PQ-QP$ need not have sense everywhere' (von Neumann, *op. cit.* p. 234). But it must, logically must, have sense everywhere in the same physical language. Where it ceases to have sense is where one physical language ends and another begins. The transition may be gradual, and the two languages may even be formally analogous at many points.

PAGE 156

1 Cf. von Neumann's (*op. cit.*) fundamental statistical formula on p. 295.

2 'Not only is the [simultaneous] measurement impossible, but so is any reasonable theoretical definition' (von Neumann, *op. cit.* p. 326).

3 So too Heisenberg considers a single event: '*Wave aspect:* Electron creates field; field acts on another electron. *Particle aspect:* Electron emits photon; photon is absorbed by another electron. Both statements describe the same event.' And again: '*Wave aspect:* Neutron creates field; field acts on proton. *Particle aspect:* Neutron emits electron plus neutrino; electron and neutrino are absorbed by proton' (*Nuclear Physics*, pp. 97–8).

PAGE 157

1 Cf. Fermi, *Elementary Particles*, Appendix III, pp. 104–5. But Temple conjectures: 'The Correspondence Principle is only a temporary

expedient which must sooner or later be replaced by a more profound study of the nature of microphysical systems' (*The General Principles of Quantum Theory*, p. 75). This is plausible in a way in which the same remark made about the uncertainty principle could not be.

PAGE 158

1 'Die Welt ist, trotz aller Ätherschwingungen, die sie durchziehen, dunkel. Eines Tages aber macht der Mensch sein sehendes Auge auf, und es wird hell' (Wittgenstein, *Phil. Inv.* p. 184).

PAGE 161

1 The fundamental difference between wave mechanics and quantum mechanics must never be obscured by proofs of their observational isomorphism, though Schrödinger once acceded to the likelihood of a statistical theory (*Naturwissenschaften*, XI (1923)). De Broglie's thesis seemed to offer him a 'classical' hope. He tried to establish a link with field-physics, particularly with classical electro-magnetic theories of the Maxwell-Hertz-Lorentz type. His ψ-waves were vibrations of the electronic charge continuously distributed over the whole wave field. Against this, Quantum Mechanics is based on discontinuity and (particularly in the case of matrix mechanics) with the rejection of unobservables. Initially it required cumbersome calculations with infinite matrices; since wave mechanics employed a familiar theory of partial differential equations it attracted physicists. But with Dirac discontinuity came into dominance.

PAGE 162

1 We are not yet discussing De Broglie, or Ψ waves, whose phase velocities are physical fictions and whose group velocities are measurable only on suitable interpretations of Ψ^2.

PAGE 164

1 '$\Delta\psi$' is often rendered '$\nabla^2\psi$'. We shall adopt the former notation. Thus:

$$\Delta\psi = \nabla^2\psi = (\partial^2/\partial x^2 + \partial^2/\partial y^2 + \partial^2/\partial z^2)\,\psi.$$

PAGE 165

1 Remember, however, that the single value condition is significant only when change in λ is so gradual that the orbit as a whole can be divided into intervals within which several equal λ lie. This condition is not always satisfied.

2 $\Psi\overline{\Psi}$, often written Ψ^2, or $|\psi(x)|^2$, corresponds to what, in § D, was called the Ψ packet. In the present notation Ψ itself would designate only an individual time-dependent *onde de phase*, e.g. $\Psi_1, \Psi_2, \Psi_3, \ldots$ Our

concern henceforth is thus with the physical interpretation of $\Psi\overline{\Psi}$, or $|\psi(x)|^2$. One can dismiss Ψ as a mathematical fiction, even in Schrödinger's theory.

PAGE 167

1 Schrödinger also provides a vivid derivation of the dipole components which determine the intensity of spectral lines: $P_{kl}^{(q)} = e\int q\psi_k\psi_l\,d\tau$. These components are identical with Heisenberg's matrix elements.

2 In particular, the wave equation for a single particle,

$$\Delta\psi_\alpha + 8\pi^2 m/h^2(E_\alpha - U)\psi_\alpha = 0$$

was generalized by analogy with the classical Hamiltonian function $H(p, q)$ into the important

$$\left\{H\left(\frac{h}{2\pi i}\frac{\partial}{\partial q}, q\right) - E\right\}\psi = 0.$$

This is now a perfectly general wave equation. Its enunciation led to the recognition of the relationships with Heisenberg's theory, and ultimately to the proof of the operational identity of the two approaches.

PAGE 168

1 *Wave Mechanics*, pp. 17, 19, 21, 53.

2 *Annal. Phys.* LXXIX (1926), 489.

3 Sommerfeld, *Wave Mechanics*, equation 24, p. 107. The material density of the electronic mass is

$$\mu_0/eS_k = h/4\pi i(\psi(\delta\overline{\psi}/\delta x_k) - \overline{\psi}(\delta\psi/\delta x_k)) - e/c\Phi_k\psi\overline{\psi}.$$

4 Schrödinger, *Wave Mechanics*, equations 52 and 53.

PAGE 169

1 For more detailed accounts of the failure of the Schrödinger interpretation of Ψ, see: Halpern and Thirring, *The New Quantum Mechanics*, chs. V, VI and X; Heitler, *Elementary Wave Mechanics*, chs. I, II and p. 64; Mott, *Wave Mechanics*, chs. I, II; Weyl, *Gruppentheorie und Quantenmechanik*.

2 Einstein, *Berlin Ber.* (1923), p. 359. Bohr, 'Das Quantenpostulat und die neuere Entwicklung der Atomistik', *Naturwissenschaften*, XVI (1928), 245.

3 Born, *Z. Phys.* XXXVII (1926), 863.

PAGE 170

1 Note the never-vanishing possibility that the particle could appear at any distance whatever from *B*.

PAGE 171

1 In other words, for more than one electron $\Psi\overline{\Psi}$ measures the probability that the particles lie in the $3N$ dimensional volume element of configuration space. Averaging over a large number of atoms this gives a mean distribution of charge corresponding to Schrödinger's *verschmiert* $\Psi\overline{\Psi}e$.

2 Cf. Jeffreys: 'What Heisenberg pointed out was that any attempt to measure the position of an electron would require the use of radiation of very short wave-lengths [incidentally, this is not at all what Heisenberg pointed out, as a glance at *Z. Phys.* XLIII (1927), 172 would reveal] it would still be possible to maintain that the co-ordinate and momentum had exact values at any instant, but that they changed discontinuously at an observation.' And even after considering whether there might not be a more fundamental question of principle involved here, Sir Harold concludes with: 'I do not think that any of the standard arguments against determinism in quantum theory are conclusive against determinism holding in the classical sense' (*Scientific Inference*, pp. 219, 220).

PAGE 172

1 Jeffreys, *Scientific Inference* (2nd ed.), pp. 215, 221.

2 Cf. Einstein, *Bericht vom Solvaykongress* (1927) (Diskussionsbemerkungen Gautheirs—Villars et Cie, Paris, 1928); Einstein, Podolsky and Rosen, *Phys. Rev.* XLVII (1933), 777.

3 De Broglie, *The Revolution in Physics*, esp. ch. 10; and cf. the note on page 217: 'J. von Neumann has proven that the probability laws of the new mechanics are incompatible with the existence of a hidden determinism, which makes it most improbable that determinism in atomic physics will be re-established in the future.' Compare this with the clearly contradictory announcement of p. 302: '1952—Revival of the deterministic interpretation of quantum processes (De Broglie, Bohm).'

4 D. Bohm, 'A Suggested Interpretation of the Quantum Theory in Terms of "Hidden" Variables', *Phys. Rev.* LXXXV (1952), 166.

PAGE 173

1 *Ibid.* p. 166.

2 *Ibid.* p. 167.

3 *Ibid.* p. 168.

4 *Ibid. loc. cit.*

5 De Broglie. *Comptes Rendus*, CLXXXIII (1926), 447; CLXXXIV (1927), 273; CLXXXV (1927), 380. Cf. also *An Introduction to the Study of Wave Mechanics*, chs. 6, 9, 10.

6 *Bericht von Solvaykongress* (1927). De Broglie's hindsight on this exchange, and indeed on the whole controversy connected with the interpretation of $|\psi(x)|^2$, does not seem to be in every way reliable (*Rev. in Physics*, pp. 231 ff.).

7 *Z. Phys.* XL (1926), 332.

8 *Op. cit.* chs. IV, VI (§ 3). 'An explanation [by hidden parameters] is incompatible with certain qualitative fundamental postulates of quantum mechanics' (p. x).

9 Bohm, *op. cit.* p. 169.

PAGE 174

1 *Ibid.* p. 171. Bohm here is making the highly debatable suggestion that quantum statistics need be no different in principle from the classical statistics of Boltzmann, as used in nineteenth-century thermodynamics and gas theory.

2 *Ibid.* p. 167.

PAGE 175

1 Professor S. E. Toulmin, Lecture to The Cambridge University Philosophy of Science Club (1955) entitled 'Priestley vs. Lavoisier'. Published in *J. History of Ideas* (April, 1957), pp. 205–20.

2 *Z. Phys.* XIII (1912), 973–7.

3 *Phil. Mag.* 1925–8.

4 *Phys. Rev.* LXXXVII (1952), 389.

INDEX

INDEX

INDEX

INDEX